NATURE NOTES

TIM DEANE

Acknowledgements

I'm grateful to the committee of the Organic Growers Alliance (OGA), and particularly to Jim Aplin, for their support of this project. I am indebted to Phil Sumption – editor of The Organic Grower – for his belief that the notes warrant being made available to a wider readership, and for all his work in bringing this book into the light of day. Thanks also to Holly Stevens for the cover, to Joanne Marr for the illustrations within and to Katie McGahey for the photographs.

I'm especially grateful to Jan – my surest critic. Two pairs of eyes are three times better than one! This book is dedicated to her.

Cover design by Holly Stevens

Illustrations by Joanne Marr (Except farm map on p31 by Tom Dawn/Jan Deane)

Photographs: Tim Deane p162, Michael Lathwaite p169 and Katie McGahey p172

Book design by Phil Sumption, Bio Communications

Published by Choir Press in conjunction with the Organic Growers Alliance

ISBN 978-1-78963-489-1

Contents

Foreword by Phil Sumption.. V

Introduction VII

1. Bees, birds, cats and clover.. 1
2. Timing is all 4
3. The Holly bears the crown... 7
4. "There's a flower that shall be mine −..................................... 10
5. Confusion, clarity, obscurity.. 12
6. The Hornet Drum... 15
7.   Pheasants, predators and penguins... 18
8. Water, weeds and consciousness 21
9. Soil associations 24
10. Chicory blues... 27
11. What's in a field name? 30
12. The Merry Month... 34
13. Eating disorders... 37
14. "Pray you, tread softly −... 40
15. Almost a vegetable 43
16. Bird business 46
17. The swallows' summer... 49
18. The night.. 52
19. Land and water... 56
20. Life chances 59
21. Spiders, from Mars? 62
22. Where are the birds?.. 65
23. Light... 68
24. Two onion kind.. 71
25. Some hardcore natural history... 74
26. B is for beaver?.. 78
27. Animal spirits 82
28. Summer skies... 85

29. Mind that tree 88

30. Hip, hip 91

31. A sense of purpose 94

32. Invasive plants 97

33. An elegant sedge 100

34. Cabbage Whites 103

35. Leeks, from salt to silt 107

36. The opposite wood 111

37. Water's ways 114

38. A bit about ivy 117

39. Stones 120

40. The turtle's voice 124

41. Nuts! 127

42. Winged life 131

43. Small lens – big world 134

44. Whatever next? 137

45. Seed sense 140

46. Plant life 144

47. While hedging 148

48. Dandelions 152

49. Adam/Had 'em 155

50. Survivors 159

Epilogue: Two valleys and a farm 162

About the author 172

Foreword

The Organic Growers Alliance is very pleased to publish this set of Nature Notes, written by Tim Deane for commercial organic growers, which we believe deserves a wider audience.

I consider myself privileged to have known Tim for many years, as a fellow organic producer, as a horticultural adviser for the Organic Conversion Information Service, run by the Organic Research Centre – Elm Farm in the 2000's and then as part of the second coming of the OGA – re-formed as the Organic Growers Alliance in 2006. The original OGA, the Organic Growers Association ran from 1981 until combining with British Organic Farmers in 1991 and later to merge into the Soil Association's Producer Services Department. The Organic Growers Alliance was set up by a group of organic growers to support and represent growers of all sizes who wish to produce their crops organically.

It was at an early meeting of the new OGA in 2007 that the subject of communications and publications was tabled, and Tim and I volunteered to produce a quarterly journal which was to become The Organic Grower. Approaching 70 issues later, it is still going strong. From the first issue Tim was able to eloquently express what it means to be a grower and 'the uniqueness of our craft.' Though Tim stepped back as editor of The Organic Grower in 2011 he has continued to contribute on a regular basis ever since. As a hugely respected pioneer of modern organic growing, and with his wife

Jan one of the first to champion the mixed organic vegetable box, Tim's technical articles are always authoritative and informative. However, Tim's Nature Notes which first appeared in *The Organic Grower* in Issue 8, Spring 2009 have also proved enduringly popular, the first columns that many growers turn to when opening the magazine.

Organic growers work with nature and their livelihood depends on it. It is that empathy, observation, and attention to detail that made Tim such a good grower. As Tim wrote in the very first *Organic Grower*. "The grower has to enter the soil in which (the plants) root as well as to live the weather in which they grow… It is the grower who best preserves that vital link of mankind with the earth and its processes. The sun's energy, photosynthesis, and the cycling of carbon – this is the basis of all life. In the growing of plants organically lies its truest human expression."

The perspective of the grower provides a unique window onto the nature of a small Devon farm on the fringes of Dartmoor to its fellow inhabitants and occasional visitors. While the purpose of the book is not to be an instructional book on the practices of organic vegetable production, it will provide some insights into the craft and the potential for organic farms to not just be oases of nature in a degraded agricultural landscape but beacons of hope for the future.

If you would like to find out more about the Organic Growers Alliance and to receive *The Organic Grower* magazine visit https://organicgrowersalliance.co.uk/

Phil Sumption, Editor The Organic Grower.

Introduction

The Nature Notes collected in this book first appeared in each quarterly number of *The Organic Grower* between 2009 and 2021. The OG is the magazine of The Organic Growers Alliance, the membership organisation for British commercial organic horticultural growers. The notes were conceived as a generally light-hearted addition to all the essential technical information which that magazine so necessarily and so ably conveys. Their title is somewhat tongue in cheek – I make no claim to being a proper naturalist! Instead they are reflections arising from a working life on the land. Their intention was to add a bit of additional colour to the grower's already colourful experience of that life, and perhaps to illuminate some of its engaging and sometimes mysterious aspects. It is hoped that the information they contain is at least roughly right, and not precisely wrong.

Looking back at them I see that there are quite a few references to the weather conditions current at the time. I make no apology for this – while the weather is a subject matter that soon goes cold (and is more appetising when served warm) it is of unavoidable importance to all growers and farmers.

I've taken the opportunity to revise some of the notes slightly so as to make their meaning clearer to a non-specialist audience, and to the same purpose have added additional explanations (in italics) at the end of a few.

Bees, birds, cats and clover

Spring 2009

If dogs are a mixed blessing on the farm and vegetable holding, then cats are a necessary evil. The evil comes not from the mere irritation of their habits – depositing guts and carcasses on the door mat, crapping in the polytunnels and so forth – but from their depredations of the bird population. To the cat, more earth-bound than it would like to be, the quick, bright movement of birds and their ability to fly is an irritant and a challenge to which only the laziest or most inept will fail to rise. Last year about this time, a pair of long-tailed tits came constantly to our windows where they winkled out skeins of spiders' silk from the little vents at the top of each casement. Anybody who has found a long-tailed tit's nest, wonderfully intricate bottle-shaped constructions often placed within the deep recesses of a bramble thicket, will know that no effort is spared in making these as luxuriously soft and warm as may be. Hence the time spent gathering the spiders' weft. Alas – our cat had them both. Opening the front door one morning a sad draught blew in pathetic little clouds of their feathers.

A little while ago, in the brief moment before I got to the off switch, Ann Widdecombe came on the radio to tell listeners that her cats were always kept in at night, and therefore could not be held responsible for slaughtering wildlife. As birds are about by day but (like Ms Widdecombe's cats) tucked up at night, I felt that this missed more than the half of it.

The necessity of cats for farmer and grower is the undoubted protection they give our stores of apples, potatoes, seeds etc. from mouse damage. For organic farmers in particular there is another, more subtle or circuitous way in which the cat compensates for its destructive habits. This has been a wonderful spring here for bumble bees. I don't remember another time when so many queens have been on the wing in March and April, intently quartering the hedge bottoms and thickets for nest sites. It was Charles Darwin who first pointed out the line that runs from cats to clover, via mice and bumble bees. Red clover and heartsease (*Viola tricolor*) are largely or wholly reliant on bumble bees for the pollination of their flowers. Field mice can destroy a very high proportion of the bees' nests. Where cats abound and kill the mice, more bees survive and so the clover sets more seed and flourishes. Lately, besides our own moggie, several neighbouring cats have taken to spending time in our fields (and perhaps also bearing some responsibility for the crap in the tunnels). If Darwin's hypothesis is right the humming and bumbling that has given us so much pleasure these past two months is explained and the cats to a small degree forgiven.

The little birds live at a faster rate than us humans. Of course they do – they have to fit a lifetime into a shorter span. It is said that they are eight times more tonally acute than we are. If bird song is recorded and played back at an eighth of its normal speed what you hear is something very much more musically complex. The skylark's song, for instance, is revealed as full of curlicues and grace notes that we would not normally hear, but which birds obviously do. So there is more to the lovely and joyous outpouring descending from that distant speck of beating wings than we could ever imagine.

We are not in skylark country here but at the other end of the spectrum, we do have ravens. Usually regarded as birds of ill omen, they strike us as cheerful and even jokey creatures. They nest on the heights above, perhaps preserving some ancestral memory of Scattor Rock which was the easternmost tor on Dartmoor until quarried away for roadstone in the 1940s. Their lifespan is long, and they can hardly live at the eight times pace of the little birds. Sonically speaking they are pigs with wings, snorting and oinking their deliberate way across the sky. Sometimes you may be lucky enough to see them fly on their backs, even corkscrewing as they go. Until recently a 16 foot pole carried the electric line from our packhouse-shed to the caravans in which we used to live. It was our cat's delight to run up this pole and sit on the top of it. One day it was up there feeling pleased with itself when a raven suddenly appeared, determinedly making its way just above the top of the adjacent overgrown hedge. The mutual shock caused the raven to apply its brakes and sheer off, up and away sideways, while the cat reared in alarm at this greater blackness than its own but clung on to its unlikely perch.

NATURE NOTE 2

Timing is all

Summer 2009

Above the din of the Massey Ferguson 165 tractor I just registered a roar of jet engines and looking up caught sight of the Red Arrows in eponymous formation heading south over the tops of the hedgerow trees. A second or two later, before quite vanishing from view, they released trails of coloured vapour – red, green, yellow and smoky blue – as (I reckoned) they passed above the village field. Two days early – the Show is not taking place until Saturday, and it is now Thursday. Not too early to take a six inch rule to the courgettes, or pick some choice plums before they split, but too early to be The Attraction.

Too late now to get rid of the docks before reseeding, for all my scuffling. The field concerned, at 4½ acres, is by far our biggest. Ploughed too deeply in June after a cut of silage, it needed July to work the roots up to the surface, but with the wet that came it was well into August before I could even venture onto it with tractor and trailer to remove the stones revealed when working down the ploughing. I should have known! Back on the vegetable ground there was little weed germination through dry spring working and early summer sowing. We had above an inch of rain on June 6th, followed by three damp days and more rain on the 10th. This induced no germination to speak of and it is easy to see why the weeds stayed where they were, given that the weather thereafter remained mostly hot and entirely dry until into July. The 6th of July brought ¾ inch. This time the rain stimulated immediate and heavy weed germination, even before more and, as it turned out, incessant rain arrived on the 11th. They know, you know! This is not the first time I've noticed that a brief punctuation of a drought, even a lot of rain will not fool more than a few weeds into appearing. They prefer to wait unto a constant supply of moisture is in the offing and then, somehow sensing their opportunity, appear in great profusion almost overnight.

Despite the miserable July our pair of spotted flycatchers brought off a second brood this year, which is one more than they managed in either 2007 or 2008. I say "our" because they are a constant of summers on this farm. For the last three years they or their predecessors have chosen to nest at the back of the Other Shed, a little above fingertip height on a ledge near the top of a corner post. This is a frequently noisy spot, just behind the tractor and link box and just above where the hoes, shovels etc are hung, but that does not worry them much. The hen sits on the eggs, her head in full view. Soon after hatching you can see the nestlings gaping over the rim of the nest and then, a day or two before they fly, the fledglings move up on to the top of the post where there is more room, more breeze and perhaps less in the way of parasites. The adults fish rather than hunt, mostly from one or two favoured perches. There is a hedge corner and gateway to the side of the shed and this is their ground. No doubt the shelter of tall vegetation combined with the open passage into the next field makes it a favoured spot for flying insects. A bare elm branch projecting from the hedge is a fine vantage point, but in the past the headstock of a rotavator parked close by has done equally well. There the bird waits, about robin size and an unremarkable chestnut brown above and whitish grey below, but noticeable for its alert and springy posture, its slightly oversize head, its frequent call (tzee-tuck-tuck) and its sudden short forays into the air. Then, moustached with insect body and wings, a quick flight back to the nest and those insistent beaks that gape from dawn to dusk.

Soon they will be gone, the first brood may have left already, fattened on the fruits of the English summer air. Recently a Swedish friend, a keen bird-watcher whose work often takes her to East Africa, came to stay. She noticed the fly-catchers (they have them in Uppsala too) and spoke of having seen them also in Uganda during the other, non-breeding half of their year. Her description of how insect life declines there in the dry season just as it does during our winters explains their months of journeying. How they ever came to make that journey in the first place (and how, among other questions, it is charted) is surely inexplicable. But their time departs in one place and arrives in another, and perhaps we can feel honoured that they choose to nest with us and leave it at that.

The Holly bears the crown

Autumn/Winter 2009

Approaching the bank that climbs steeply from the flat space alongside the brook I registered a congregation of chattering small birds in the uppermost branches of a big old ash tree that stands at the top of it. Going up and getting to its other side, so that the sun was no longer in my eyes, and gazing upward I thought I might see a roosting owl, caught in the bright light and at the mercy of its small and insistent tormentors. But there was nothing of that sort, just the sunlight glinting on the silvery bark and the fat pendulous twigs etched against the limpid sky, and soon enough – whatever had been its cause – the gathering dispersed. Alongside the ash, growing out of the low stone hedge, a field maple (an altogether neater and more demure tree) reaches up to perhaps half its height. The ash is quite bare of leaf of course, even in this protracted autumn – it now being the second week of December. The maple still holds on to several of its prettily lobed leaves, which are a lemony yellow. Beneath both trees and between them is a holly, not much more than a bush. Last time I looked it bore a good crop of berries. Now there is one left.

One of the several things on our list when we were looking land all those years ago, almost of all of which were satisfied by this farm, was a good number of holly trees. From a quite early age the chief pleasure of Christmas for me has been to bring the outdoors indoors. Short of space when we lived crammed in caravans my ambitions had to be fulfilled in the packhouse, with the advantage that the often frigid air kept the greenery as fresh as when it was cut for the 12 days (actually 13 by my reckoning) of Christmas.

You can see what sort of crop of berries there'll be when the fruits start to swell between the fading petals of the flowers in May. By early October most of them have turned to their ultimate jewel-like blood redness, and in a good year they do much to add to the general jollity of the autumn hedge and brake – outshining the haws and only rivalled for redness by the rose hips, which however seldom come close to the holly in the way of massed brilliance. By November the favoured trees are already being stripped by birds, the ground beneath littered with the ones they've dropped. By December the migrant thrushes – fieldfares etc – and south or westward moving blackbirds are making serious inroads into them, so that by Christmas Eve in some years there are few berries left. Since we have been here the point at which most of the trees have been relieved of their fruit has come inexorably if unevenly earlier. I used to think that cold weather, here or at a distance, was what governed the birds' harvesting, but now I am pretty sure that the ripeness or otherwise of the fruit is the chief or only factor. This earlier harvest is thus another sign, though I would rather it wasn't, that change is coming quickly to the world around us.

I used to sometimes shove a produce net over some especially choice branches (not that easy on account of the prickles) to ensure some well berried bits for myself. I don't bother with that any longer. The last several years have all been good ones for holly berries and the holly is diverse enough as a species and our stock of holly trees great enough that there is always something to be had on Christmas Eve. In fact we have a few specimens whose fruits are usually still tinged with yellow at the solstice, and the birds never take them at that stage. The main thing is to have a decent piece for the pudding, and a few other ritual settings. The full glory of the holly is the blood red against the rich green, but the leaves alone in their profundity of colour and the faceting of their mirrored surfaces, gathering and projecting back the light, bring something to the indoors which is both earthly and celestial. Thus, do we mark the turning of the year and honour what has passed and what is yet to come.

"There's a flower that shall be mine –

Winter/Spring 2010

'tis the little celandine". So wrote Wordsworth in one of his many lesser poems. Today is March 7th, St Perpetua's Day. This is supposedly the day the celandines come out, though I feel this is more likely to be nineteenth century whimsy than real folklore. Often, sometimes for years in succession, the Westcountry winter is a halting, non-descript thing and we can seem to go from autumn into almost-spring with barely a glance of it. Then the celandine are out en masse well before the end of February. There's always the odd intemperate individual. I even found one attempting to flower right at the start of this year in a brief lull between the gripping frosts. There are one or two to be seen now, but they are poor pinched things, their petals bleached and curled by the continuing frosts (minus 4°C again last night). I doubt that St Perpetua, with (in beatification) her head restored to her body, would recognise them as honouring her. I don't remember a year when they haven't been making a show by now even when we do get a winter, but this year is different. We'll have to wait awhile yet before those massed stars shine out from the hedge bottoms, challenging the sun itself for brightness.

The snowdrops go on and on. A bit ragged now after six weeks in flower but still they light up otherwise overlooked corners, in places advancing out into the open ground, where so far there is little more than a suggestion of the new year's growth. Below the orchard by the house there is a sunken lane. From that vantage point drifts of snowdrops spread out at eye level beneath bare boughs of sycamore and hazel, a sheen of glossy whiteness that here and there drips down the stony retaining bank to gather into little bright pools of light in the shade at its foot.

Meanwhile, up in the heavens, the advancing year will not be denied. As Orion the hunter and his dog sink westward with the winter so Bootes climbs the eastward sky. He is the ploughman who with the lengthening days puts his hand to the Plough. I like to think of Arcturus, among the brightest of stars, as his eye reddened by the keen March winds, though I suspect that, anatomically speaking, it is more likely to represent a chilblain on his toes. Either way, in this age of tractor cabs, the metaphor does not have quite the force that once it did.

NATURE NOTE 5

Confusion, clarity, obscurity

Summer 2010

Each spring or early summer soon after we lay our bit of black plastic for the outdoor cucurbits there'll be some mayflies and the like, fluttering above the strip between frequent dabbing descents. These I presume are to do with the laying of eggs, though I cannot say whether they are just testing what they take in its reflective darkness to be water or are actually doing so. I told Joan Loraine about this while she was here judging us for the Loraine Award (for nature conservation and organic farming) in 1997. "I think that's depressing" she said, and who could argue with that? A few weeks back there was a large blue dragonfly hawking back and forth, over the shiny strip. This may have been because it too mistook the plastic for water, or because its prey did, or both. I saw it several times, but no longer now that the sheen is disappearing under the energetic expansion of the squashes etc. Do black automobiles cause the same confusion? It is not to be wondered at – the compound eye of insects is (it seems) wonderfully adept at registering movement but is not designed for the appreciation of landscape or its composition. We needn't feel superior – the mayfly (which lives all its life in streams and rivers, until its winged epiphany) knows more about water and air than we ever will and can't be faulted for knowing nothing of the properties of petrochemicals.

What does the swallow see, I wonder, as the world streams past it in shifting planes of land and sky? And what have they seen, between here and the Indian Ocean edge of South Africa? It is a pleasure to have them back in our packhouse-shed after a gap of ten years. Today the young, with great excitement, flew for the first time and then gathered again to perch on the steel principal close to their nest. A pair has refurbished one of three old nests. This one is stuck, as the remains of the others are, to a galvanised purlin but with the additional anchorage given by the side of a strip light fixed to its underside (at this time of year we can manage without switching it on). We can peer into it from the railed edge of the storage loft, at eye level and only just beyond reach. I saw the cat there, on the top rail and waving its paw in the direction of the nest. Since then we've kept the hatch shut. When it passes by the front of the shed the parent birds swoop down and chitter at it, which it affects not to notice. They are not especially rapid fliers as small birds go, but their mastery of the air is complete and if there is anything more pleasant to watch on a summer's day than swallows and martins at their sport, I don't know what it is. An old house that we lived in had a dozen pairs of house martins nesting under the back eaves. From the bathroom you could listen to their cheerful conversations just above the open window. OGA member R. Hall has a tall and venerable stone and cob barn on his farm that has the unique distinction of being home not just to swallows and house martins but swifts as well and, if you can credit it, sand martins too – you can see their holes! Thus our Roddy is guardian of what ought to be a national treasure.

An unexpected plant has appeared in one of the polytunnels, just beyond the bit of staging where I do my plant-raising. I couldn't recognise it at seedling stage but before long, on account of its long and narrow leaves, I thought it might turn out to be a weasel's snout (or lesser snapdragon, *Misopates orontium*) and so it has. A once not very common and now perhaps quite rare weed of arable ground, I've seen it just twice in our 25 years here. Then, in the shade of vegetables, it's been a rather drab and meagre thing. This one on the other hand managed to come up right beside a drip line and has been allowed to grow without competition. Now, early July, it is knee high and sparsely decked with delicately formed pink blossoms, like miniature snapdragon flowers.

The strangest plant I ever saw in a polytunnel was what I took to be greater dodder (*Custuca europea*). I'd never seen it before. According to the books it is a rare parasite of nettles, hops and maybe thistles. We have plenty of all three, but this dodder was in the leek seed bed and apparently parasitising a leek. Lesser dodder is quite common in heathy places, festooning furze bushes with tangles of blood red threads and packed clusters of tiny pinkish white flowers. It, or a closely related species, used to be a common pest on clover crops until improved seed cleaning techniques did away with it. Dodders are entirely parasitic – the seedling advances along the ground until it finds or fails to find a suitable host. If the former it then spirals up around its stem in both loose and tight coils. Where the two plants touch closely the dodder sends out suckers that penetrate the host's cell structure (all this from a seed, without any photosynthesising!). Through these the dodder draws all its sustenance and water, sometimes to the extent of killing its victim. It has no more than sparse vestigial leaves, like scales, and no chlorophyll of its own – just threadlike stems and masses of flowers. Though the one in the tunnel never got to flowering stage. What could I do? It was inextricably mixed up in the leeks' seedbed, and their time for planting had come.

NATURE NOTE 6

The Hornet Drum

Winter 2010

Is it just me, or are there really less insects about than there used to be? I don't mean the pest species of course – like pigeons and rats and the rest of our fellow travellers, they are doing well enough – but the benign hum of myriad arthropod life seemed muted this summer past. It wasn't a bad summer here for crops or (you'd think) for wildlife, unlike in some parts, but perhaps the three excessively wet ones that preceded it are the cause. Even when harvesting clinched melons I never saw a wasp and barely a hornet. After several years the latter didn't this time around nest in their familiar hollow in a decaying ash tree that straddles the hedge between two of our vegetable fields. Hornets are muscular killers in the insect world but they don't usually bother us, are handsome and prominent. I like to see them but I also like to know where their nest is – I'd sooner not stumble on it unawares. By late summer all their hope of posterity is vested in the over-wintering queens that they've reared and tended, and the colonies break up, the workers now on the holiday of a lifetime. In the evenings they come to the lighted windows and then, if a window is open, they do make a nuisance of themselves – even this year we had a few. A strong wind stirs them up. One windy October night four came in, in steady succession. They always, each one of them, do the same thing, perform the same manoeuvres. First a straight line, a bee-line if you like, from the window to the paper light-shade that hangs above and a little to one side of the kitchen table. The shade resembles nothing so much as a gargantuan version of the first founding cell of the hornets' own nest – white, papery, globular and curiously light to the hand, but its magnetic attraction to them is the light within it. It is as much to them as to any moth, an obsession not apparently shared by other vespids.

The hornet has a hard body. It also has a vigorous and headlong flight. The consequence is that it's noisy, very, both in its wing movements and, when mazed by the light, in its crashing against solid objects. I have known them knock themselves out against window glass, sprawling on the sill or ground beneath for a while until they recover sufficiently to have another go. Once in and at the light they fly and dance around its surface, bashing the tissue paper with their heads and constantly scuffling with their feet while the intense rapidity of their wing movements fills out the rest of this manic drum and bass with an insistent tone, simultaneously roaring and needling. Soon enough they find their way into the globe, giving the sound an added resonance while also dislodging any accumulated dust and debris that has resisted earlier drummings. This, in contrast to its dislodger, floats gently floor-ward. After a while, sated for a minute, the hornet climbs onto the rim of the upper opening from where it casts an exaggerated shadow onto the ceiling. The sound dies away while its body palpitates restlessly. Only the sound – that by now has got into your head – somehow goes on, as if a physical presence embedded in your synapses. Then perhaps another bout of drumming, and maybe another little rest, before the inevitable next stage. This consists of a series of erratic orbits of the area that lies between the light-shade and the dimmer end of the room, performed in a head-up, dipping flight at maximum whirring speed, each dip punctuated by a bash of the head against the plaster.

At some point even hornets get tired, or sore. If we are lucky this happens sooner rather than later. Then the beast (for such it has now become to us) goes to ground, or at least to some dark surface. For good reason it is always a dark surface – you must keep your eyes open to mark the spot! Now is the time to strike, quickly, before it revives, returns to the light-shade and sets the whole mad ballet in motion once more. A hard surface, a magazine – between the two brought forcibly together it meets its end. There is relief in this, but no pleasure. The creature will soon die anyhow, but this is an ignoble end. It leaves behind it a diminishing trail of sound.

NATURE NOTE 7

Pheasants, predators and penguins

Spring 2011

There's a plague of pheasants locally which has become something of a cause célèbre, with a television programme (which I failed to see) devoted to the problem. The reason for it is what can only be described as the industrialisation of pheasant shooting; the result – that the valley woods between the beauty spots of Steps Bridge and Clifford Bridge begin to resemble over-stocked chicken runs. Of course the birds are fed close to the one narrow public road that meanders up the valley and so, equally predictably, they hang around along its length and in their scores considerably impede the motorist's progress. More seriously the pheasants are altering the ecology of the woodlands, for instance by gobbling up the larvae of butterflies and other insects.

I worked on the home farm of an estate for a few years. Much to the disgruntlement of the farm manager (who only had me to manage) I was required to turn out as a beater for the nine driven shoots each season. His stern Methodist radicalism made him an objector to the whole business, but I enjoyed it as a day out rambling around the side valleys and woods of the upper Tamar estuary with plenty of different company and often enough a good laugh at the expense of "the guns". On one occasion I even had the thrill of shot whistling by just above my head as a woodcock (a red rag to a bull for a gun) took off from in front of my feet. But generally the thing was kept within bounds. Certainly, there were plenty of pheasants around – two full-time keepers saw to that, but they also saw that the birds stayed by and large where they were put. Although the guns paid handsomely for their sport, the object was to satisfy the squire's fondness for the pastime, not to make money – which even with pheasant feed put through the farm account it probably didn't. Meanwhile the farm staff, Ivor and I, had to break into the shoot's store to get hold of some netting so we could fold off the kale to the sheep. But that's another story.

This current story concerns twenty-eight shoots in each short season, the integrity of a SSSI nature reserve severely threatened, and locals and visitors having spent shotgun pellets and even dead pheasants falling on them. I wouldn't say any of those birds are finding their way here (dead or alive), but a neighbour has caught the bug and is releasing his own to give himself and his mates some extra sport. He grew a few acres of swede and kale just a field away across our boundary. The lamb's tongue (fat hen) which had completely obscured the crop by August just added to its value as pheasant habitat. Before the winter was far advanced a dozen hens and several cocks had moved in with us – lurking in the sweetcorn stubble, hanging round the bird table and scratching about in the bullocks' bedding. Now that it is spring and the doors are open they are making a nuisance of themselves in the tunnels. I don't mind the odd pheasant about the place. For one thing there is always a casserole if you need it. But pheasants and polytunnels, that's not a good mix. Last year there were four or five that ripped the lower leaves of the aubergines to rags, possibly because they fancied the aphids. In the end I did for the pheasants and the ladybirds did for the aphids. There's a few holes in the plastic to show for it, and I fear that this year there will be more still.

Examining the extent of the wreckage of the brassica crop, once the weather had warmed up, I found on one cabbage both a small grey slug and a hoverfly larva. The slug had perhaps moved into some sort of refuge for the coldest of the weather, though I dare say they can take a bit of frost, but the wee hoverfly larva must have sat it out in the frozen cabbage. Well the cabbage survived after a fashion, which is more than any of the broccoli managed, and the larva seemed quite fit. I couldn't see any aphids or other food source, but no doubt it has better senses for the job than I do. We tend to think that, because these larval insects are soft and squishy, they would surely freeze to death. Such anthropomorphic sympathies are entirely misplaced. The onset of cold induces a raised level of sugars in their bodily fluids, and this acts as an antifreeze. There are a few insect species that spend the winter alive and outside, from those startlingly green caterpillars you find in various crops to the winter gnats that look too feeble to weather more than a light drizzle or moderate breeze, and they are all better equipped for it than you or I. Lawrence Hills suggested that hard winters might favour pest species by limiting the activity of the predators, but I've never been able to work that one out.

You may have seen a news story – tagging the flippers of king penguins significantly (16%) limits their chance of survival and ability to raise chicks (39%). Thus not only have we been killing penguins, the data we've collected in doing so is of no value. It only applies to penguins handicapped by having had their flippers tagged. You thought imperialism was dead? It's fine to look at the world, it's good to observe its life, but there's a limit. D.H. Lawrence said that to know a living thing is to kill it. We tag farm animals and we tag criminals, but the wild and free are not ours to own or to know.

Lawrence Hills – 1911 to 1991, author and founder of the Henry Doubleday Research Centre (now Garden Organic) which became the largest body of organic gardeners in the world. His was a formative influence on many commercial organic growers.

Water, weeds and consciousness

Summer 2011

One thing you can do this weather with a good expectation of success is to go and kill some weeds. But do you have any weeds? A now rare south-westerly brought us a bit over half an inch of rain on May 7th and 8th and as the wind then stayed in the south west and more rain was forecast it did seem that the ten week drought might have ended. But looking at the ground two or three days later there was barely a weed seedling to be seen, and from this I concluded (correctly as it turned out) that we'd had all the rain we were going to get for at least another few weeks.

It had been so dry right through the period of spring cultivations that it was no surprise that while some weed had come up there was not that enveloping flush of germination invariably produced by the combination of spring warmth and sufficient soil moisture. The weed was there of course, it always is, but it wasn't giving us the chance to much reduce its further potential. Had that rain in early May presaged the start of a wet spell then mass emergence would have followed within a couple of days. Instead a scatter of seedlings appeared over the next week to ten days and that was it. I've noticed this phenomenon (or lack of phenomenon) many times before. I remember, for instance, a fiercely dry summer punctuated by one welcome lashing of thunder rain. The next day I measured 9 inches of growth on a stem of buttercup squash between morning and dusk. The plant world was thrumming with new-found vigour and gaiety, but no new weeds appeared, the drought set in again and soon the ground was just as dry as it had been before the rain.

How weed seeds know when to germinate and when not, I have no idea (of course!). It's enough to know that they do have this intelligence. It shows itself too in their ability to distinguish between water provided by irrigation and precipitation from the heavens. I'm not saying that irrigation never induces weed seed to germinate, but I have proved (to my own satisfaction at any rate) that watering a bit of ground so as to get a flush of weed out of the way before planting does not have the intended effect. For the most part the weed is not fooled. Instead – you plant your crop, sooner or later it rains properly and then up come the weeds in their hundreds of thousands. Here you can appreciate the differences between rainfall and applied water – temperature being the most obvious one. Atmospheric nitrogen brought to ground in rainwater may also be a relevant factor, the particular rhythm of rainfall perhaps another.

None of this ought to be that surprising. Just because plants don't move around doesn't make them any less alive or less responsive to the world around them than are animals. What is odd though is that you can take your seed drill out into the field and sow this or that crop into conditions that the weeds reject as being unsuitable for growth, and given the merest hint of moisture at the bottom of the drill the crop comes up, albeit slowly and perhaps fitfully. Has the domestication of the plant altered its nature to the extent that its seed forgets inherent understanding and instead responds to our bidding? What if, instead of drilling (say) salad onions, you filled the hopper with chickweed seed – would it come up, just because you had drilled it? I've had enough of chickweed not to wish to undertake that experiment.

And then there is protected cropping, where it never rains but where, despite that, weeds have no trouble germinating.

Biodynamic understanding attributes plants with a consciousness that exists on the level of the cosmos, which would go some way to explaining the behaviour of weed seed. No doubt this idea is laughable to those who view the world from a strictly rational, reductionist standpoint. But when you consider that the functions of plant life are externalised while those of animals are internalised you may concede that the idea is not so fanciful after all. In form and purpose there is little difference between plant roots and the villi of the animal's small intestines. Thus the soil can be seen as the plant's (external) digestive organ. Similarly the sex life of plants takes place al fresco – they let it all hang out. Underlying this fundamental divide is the fact that plants take in carbon, which is hard and stiffening and roots them to the spot so that the world must come to them, whereas animals use the oxygen which is made available by this process, and which is light and quick and designed for movement. We have brains which we know about and a consciousness which we don't, but imagine is to be found within us somewhere even though we can't put a finger on it. Are we then to deny its existence in the world of plants?

A friend, at that time a student of psychology, was weeding with us one summer – long before search engines. She was trying to decide on a subject for her dissertation. I suggested plant psychology, but she rejected this in favour of something simpler to do with laboratory rats (or was it pigeons?). Perhaps her being a vegan had something to do with it. A glimpse at Google now tells of over 27 million "results" for plant psychology. Did she miss the boat?

Soil associations

Autumn 2011

Caught in congestion on one of England's great arterial motorways I was interested to spot, through the car's side window, a neat illustration of the process of soil formation. Against the concrete crash barrier of the central reservation a lone sow thistle had found itself enough detritus and root space to push up a flowering stem. Around its base dust and fragments of straw, shed by lorries trundling from the arable east to the pastoral west, were beginning to form a small windswept mound of mineral and organic matter. Left alone until next summer this might support the beginnings of a plant community, trapping more dust and adding more organic matter through the alchemy of photosynthesis (straws on the wind are not to be relied on). Meanwhile plant roots and their accompanying microflora and fungi will chisel away at impervious mineral surfaces, creating by minuscule increments the medium in which they themselves can flourish. On and in this, entirely supported by the plant, is founded an animal kingdom of wonderful complexity. Slowly, slowly the years and centuries pass and that one sowthistle, which found living space in a crack between two planes of concrete, has been succeeded by a whole ecosystem.

Of course it would take an ice age, an inundation or tectonic hyperactivity to really shift that motorway but give it a while less than that and there will be birch trees there. This is not, I imagine, what is in the mind of the body responsible for erecting boards hopefully proclaiming "The National Forest" beside the highway that leads to the East Midlands. And anyway, the Highways Agency will soon put a stop to any such pedological progression, for the time being.

It's been said that the English countryside is what exists between England's main roads. This is true in the sense that the majority of the population for most of the time experiences the countryside from the viewpoint of a travelling vehicle. From a motorway this is not to experience it at all. The country lies beyond a barrier, framed as often as not by overpasses, gantries and power lines. It is contained, sanitised by exclusion, more or less lifeless, and from that vantage point – unknown and unknowable. Here and there our trunk roads do still touch the fields and woods they pass through. Now in high summer there are places where the trimmer hasn't yet been and the bracken hangs down from the bank and brushes the passing traffic. And there are trees, lifting strong limbs above the carriageway so that the light comes down glancing and dappled by leaves. Here is illumination which is neither digital nor analogue but the thing itself – light through the leaf, which is the beginning of our existence.

Where plant and light meet organic matter is formed. Carbon fixation. Organic matter. Everything that is or was alive is founded on carbon. "Derived from living matter; containing carbon" – that's what the word 'organic' means. Do we have a problem here? I was surprised to read, in Helen Browning's responses to readers' questions printed in the last Organic Grower, that there is a 'real debate' over the 'O' word and that there is some problem between it and 'policy makers and consumers'. It had never previously occurred to me that the word 'organic' might not be spoken in full for fear of – well, **for fear of what?** Oh, I see – for fear of its lack of 'sex appeal'. This apparently makes it hard to turn into the wind and be proud of organic. Well, if you put it like that I'm not sure I'm proud of 'organic' , which, after all, is just a word. But organic farming, organic growing, organic system – what is there not to be proud of? Surprised? I thought I'd stumbled on some chink in my reality! Given its history and its purpose, sex appeal is about the last thing you would associate with the Soil Association.

Helen would like to see more stories about the life and doings of organic producers in the public domain. That's not an unworthy aim, but at root organic farming is not about personalities or products. Personalities have a place of course and without products it is nothing, but if you elevate these to be its prime and only visible purpose then you have just one consumable commodity among many and you will be judged accordingly against the fevers and prejudices of the market. Rather, organic farming is an exposition of life and the nature of creation. It starts from an understanding of the soil and its processes, that other realm – hidden and obscure – that supports our realm of light and knowable activity. Admittedly the soil is not a subject that attracts much interest among the general public (though with the wonders of microscopic photography who knows?) but that is no reason to abjure it for the idols of commerce. Going along with eternal processes, building on them where possible and avoiding their disruption at all times – I can see no shame in this. The countryside, where our food is produced, may be remote in foreign fields, but nature, life itself, is squeezed into the motorway next to us. Somewhere around here are enough truths, never mind stories, to tell what 'organic' is really about.

Soil Association/Helen Browning – founded in 1946 the SA is a charity the purpose of which is to explore and promote "a fuller understanding of the vital relationship between soil, plant, animal and man". Helen Browning has been its chief executive since 2011. Soil Association Certification is its subsidiary arms-length organic certification body. Probably the majority of OG readers are its clients, though other certifiers are available.

Chicory blues

Winter 2011

Early May is soon enough here to be sowing grass seed, and then you want moisture in the ground and the idea that there may be a bit more on the way. The late winter and early spring had been so dry, surely May would bring rain? There was a little, but I wasn't confident. Given what followed it was probably just as well that I left it until the beginning of September, by which time the season was apparently heading for an inglorious exit. The seeds grew mightily – there was a bit of rain, and then that unlooked-for Indian summer. In October I topped them and would do so again, only it is now too wet! Experience suggests that the clover will come into its own next summer. Just now it is hard to spot beneath the rampant grass growth, the again over-topping weeds (the preceding vegetables were fairly clean – where does all this shepherds purse come from?) and a good deal too much chicory.

Including agricultural chicory in a ley used to be a bit of a hit or miss affair. Sown too late in the year or topped too soon it would fail to keep up with the rest of the seeds mixture and then there'd be little of it to see. One year most of the chicory seed provided in the mix by our merchant turned out to be radicchio, which looked quaint but rather missed the point of the exercise. Usually though there was something to show for it, enough to make it a valued part of our fertility-building break. I used to tell OCIS clients that it had all the advantages of a dock – for soil structuring and nutrient recycling – with none of that plant's well-known drawbacks. What I hadn't yet realised was that the introduction of a New Zealand bred variety called Puna had rewritten the equation. It's a great chicory – quick to establish, vigorous and productive of leaf and root. However, when it came to working in the ley prior to the next round of vegetable cropping, it turned out that the Puna was not going to let go lightly. Instead it showed a dock-like reluctance to go at all. Its great fat roots, not dissimilar to parsnips, have tremendous powers of regrowth making it a big nuisance in subsequent crops and a particular embarrassment when those crops are under covers, which the stiff chicory stems then push up.

Our leys stay down longer now that their main purpose is feeding cattle rather than straightforward fertility building. It does seem that after four or five years, rather than two, the chicory is easier to kill off. The bullocks certainly like it well enough and its vermifugal properties may be a factor in us never so far having found it necessary to use any wormers in the six years that we've had our own stock. All the same I don't want too much of it because it is not the ideal material for ensilage. It tends to run up to seed while the fields are shut up for the second cut, and then the resulting product is full of inedible stalks which go on to make a considerable nuisance of themselves at the feed barrier. So one way and another there is reason to include less in the mix than the standard half kilo per acre. This time around I specified half that and that is what we were charged for, but as it is hardly possible to put two fingers down in the field without touching at least one chicory plant the invoice did not reflect the reality of what was actually put in the bag.

Chicory is reckoned a native plant but is infrequent and usually confined to alkaline soils (ours are naturally acid). One summer an elderly but spry botanist came down our lane, having been commissioned by the Soil Association to survey wildflowers on organic farms. He was greatly taken with the stray chicory which was in flower here and there, and told us that it made no difference to him that we had introduced it.

Sometimes on box days we would pick a few flowering chicory stems and put them in the packhouse for the day, shoved into a jug of water. The flowers last hardly any time at all, but while they do the washed-out sky blue of their blunted daisy petals is balmy and soothing, I would say even angelic. I can't think of another flower that approaches them in colour. It may be to the credit of organic farming that there is now a roadside stand of chicory on the outskirts of our neighbouring village, seeing as the land round about has been registered organic for several years. Speaking of angels, the Annunciation of that farm's conversion was a cereal field that was ninety percent scarlet-crimson with poppies, to the wonderment or perhaps (in some cases) derision of the locals.

Probably there is no such thing as an ordinary growing season, but some are more extraordinary than others. When I sowed that ley on September 2nd a squash plant came up just where the sown area meets the headland track. It now (November 15th) has five leaves on a short stem and the axial buds look ready to burst into rambling growth.

Growers use agricultural chicory during the fertility-building phase of their crop rotations for its value in breaking up soil compaction at depth, and for its ability to bring back nutrients that have washed down into the soil profile beyond the reach of shallower rooting plants. Radicchio (Italian red chicory) is a short-lived salad plant.

OCIS – the Organic Conversion Information Service – was wholly funded by the government and ran from 1996 to 2008. It enabled land-holders (large and quite small) to receive free advisory visits and subsequent reports assessing the suitability of holdings for organic conversion.

NATURE NOTE 11

What's in a field name?

Spring 2012

Not for the first time someone told me the other day that they were surprised to realise that individual fields have names. On one level it is no great wonder that people unconnected with agriculture would think that a field could be nameless. On another, a moment's thought would show that everything under the sun and elsewhere, tangible or intangible, has a name – otherwise how could we hold it in our mind or make reference to it? If something is nameless then so far as human perception goes it is beyond our ken and effectively does not exist. Of course there is much that is beyond our ken and in the realm of the nameless. As the Daoist motto has it –'endless the series of things without name on the way back to where there is nothing'. But once within our perception it must have a name. Even craters on the moon have names, and so do all our fields.

I worked on a farm where most of the field names were lost. It was an estate farm and the records had perished in a fire. Previously tenanted land had been taken in hand, which is why I was taken on, the manager now having more land to manage and thus the need of a worker to help him with the work. He was an intelligent man but sometimes he had a struggle to explain exactly which field it was that he was talking about. That's how I ended up ploughing the wrong field – all because we had lost the names. Fortunately he took it with better humour than he did some of my other perceived failings.

Our own farm had been pretty much abandoned for years when we came upon it. One field was named in the deeds, but there was no one to tell us what the other eight, and two orchards, had been called. Early on I went to the county record office to check out the tithe map. These were made for the majority of the parishes in England and Wales (those which had not been affected by parliamentary enclosures) following the Tithe Commutation Act of 1836 which enabled payments formerly made in agricultural produce to be converted to a cash figure. The maps and accompanying registers show every field in the parish and give their names and the names of their owners and occupiers. For some parishes, but alas not ours, they even tell you what crop was being grown at the time.

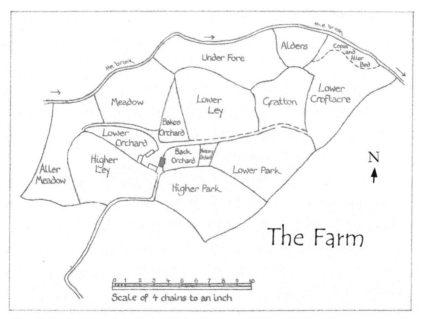

Based on the Tithe Map of 1841

Most of our field names turned out to be run-of-the-mill – Higher Park, Lower Park, Higher Ley, Lower Ley – names found on many farms. More expressive is Under Fore. At three acres our biggest field, it occupies some of the slope down to the brook, its upper boundary a sinuous hedgebank more or less marking the edge of the level ground above. Less explicable was Gratton. Then I read somewhere that this is a frequent field name in Devon and means 'burnt field' (the etymology of 'Gratton' meaning 'burnt' escapes me). In what circumstances might it have been burnt? Later I learnt that there was an immemorial method of preparing grassland for cultivation known as beat-burning. Particularly associated with this county and elsewhere called 'denshiring' (i.e. Devonshiring), it consisted of paring off the turf with a plough, spade or beat-axe. The turf, once dry, was gathered into heaps and burnt, after which the ashes were spread back over the ground. The result was a greatly increased supply of soluble nutrients to the following crop and, if overused, a serious loss of organic matter. Perhaps if we'd applied it here it would have done for our wireworm problem. Be that as it may, was Gratton a field where beat-burning was practiced? Another explanation occurs to me. At an acre and a half it's the smallest field on the farm. The name could be a play on this – Great 'un, although if you're speaking Devon the word is 'girt', so in that case it ought to be Girton.

There are few dead straight lines on the farm map, and none that are parallel, but almost all the fields manage to have corners where different alignments of hedgebanks meet at an angle. Later we added an adjoining field – Lower Waterleats – which is oval and thus without corners. On the map it stands out from the informal polygons around it, and it's striking that the several hedges that run up to it all terminate at its boundary. On this account it must be an ancient landscape feature, predating the centuries-old fields that surround it – perhaps a piece of managed woodland dating back to the distant time of open fields with rough grazing beyond which later was enclosed to form this farm.

The name 'Waterleats' speaks of another former farming practice – water meadows. The leats have gone now, cultivated out years ago, but the little stream that fed them is still there in its shallow cut at the foot of the hedgebank. Most of our fields were irrigated in the same way, through shallow channels arranged so that water led in would spill over and seep down from one to another before being finally drained away. Even on ground we haven't disturbed there is little sign of them now, except for two that are cut into the slope above the brook where they make prominent ledges as they follow the contours – as if they had once carried the tracks of a light railway. This part of the system was still in use in the 1930s, but the generation that had memory of the practice is gone now and the water meadows of Devon are all but forgotten. Like those further east some date back to the 16th and 17th centuries but are different from the better-known Wessex variety in that instead of flooding level ground they utilised water moving over a gradient, sometimes gentle but typically steep. This sort were known as Catch-work Meadows. Unlike the Watered Meadows that used alkaline river water these relied on springs and short streams, and as this water springs from impermeable rocks it is acidic and carries negligible amounts of nutrients. Nor could they supply water in a dry time since the springs that fed them are then reduced to a trickle. Charging the ground with water ready for the growing season may have been useful, but a soil moisture deficit in winter is a rare occurrence here. The main purpose was to raise the soil temperature so that grass grew even when the air was cold, an effect that can be seen as a brighter green where spring water breaks the surface of a pasture in winter. A lot of work which has now been replaced with bags of chemical nitrogen, but maybe their time will come again!

The Merry Month

Summer 2012

Sometimes, for instance in press reports referring to some new technological use that's been found for it, you see straw described as a waste product. This is likely to get a rise out of a livestock farmer. We pay around £75/tonne for the stuff and through the winter use about half a ton a week to keep our small herd of cattle adequately bedded. There is nothing about it that is waste. I wouldn't even call it a resource, a word that implies some human ownership. Of course what it really is, is land – seeing as that's where it's come from. The best way of getting it back there is via winter yarded cattle, so that its unyielding cellulose is bruised and broken by their hooves and well mixed with their dung.

The cattle are out now, continuing to do what their complex alimentary systems and simple minds best fit them for, which is to be either actively eating or passively ruminating and producing a lot of dung while they're at it. This spring event, when the beasts move from winter to summer quarters, and its corresponding time in autumn when they return to the farmstead, are the two poles of the pastoral farmer's year. Thus May Eve/May Day and Halloween/All Souls have been celebrated since the earliest days of livestock husbandry.

The spring turnout is a moment of liberation for all concerned. Bright-eyed the cattle sniff the grass, take a bite here and a bite there. Too excited yet to quite settle to it they kick up their heels and ramble around for a bit, butting each other in playful spirit, before returning to the feast at their feet. Meanwhile the farmer is freed from the daily chores of feed and bedding, and, these days, from the shameful mess of silage wrap. Instead he can cast an eye over the stock while enjoying the flowers at his feet and the birdsong in his ears, and all the new-minted glory of the spring landscape. It was though a moment not without misgivings on my part, because at last April was making up its mind to be wet and who knows what weather might be coming, so odd and strange had it all been up to then. Misgiving soon turned to grief. Six inches of rain fell during the last week of April. As much grass was spoilt by trampling feet as got eaten, and the pummelling and poaching of the ground will set back regrowth for the whole season.

Not a great time for vegetable growing either. A spring line opened up across the centre of the small piece that I'm still working, and it was back to square one. All the same, despite drowned vegetable beds and ruinous pastures, there was something wonderful about that deluge. It had been so dry for so long that something dramatic was needed. Through the winter the springs had run without conviction, the more ephemeral of them had long since dried up and a fearful drought looked certain. Then, almost overnight, fresh springs were bursting out everywhere, including the tiresome one which marks rare and exceptional high water here by bubbling up in one back corner of the packhouse shed before finding its way out through a hole in the wall near the opposite front corner.

Such minor irritations aside, this farm takes on a special life and gaiety with the running of water. The brook along the long bottom boundary, though levelling out after its steep fall from the plateau above, runs swiftly over a series of small falls and cascades, interspersed with pools and quieter stretches. Now well charged, its insistent sound (somewhere between a roar and a purr) underlies the day and more so the night, the note rising and falling according to the wind and to your vantage point. Its little tributaries, which define all but a couple of hundred yards of the farm's other boundaries and which die away almost to nothing in most summers, rush or tinkle over black stones, and glint where the light touches the current through enveloping vegetation.

What's been really wonderful is the way the headlong charge of early spring came almost to a standstill in the cold and wet of April so that there were still daffodils at Easter, which had seemed impossible in March. The older I get the more I want to hang on to this season, especially to what is often not much more than a moment of it but which this time around has lingered deliciously – the cusp of April and May, when the oak is golden in its newness and the ash just delicately bursting forth from its thick mat-black buds, and all else arrayed in infinite shades of new-minted green. Then light is immanent in the trees themselves. On sunless days they bestow this illumination on the world around them.

Seldom does the hawthorn, the flower of May, bloom by May Day on the late soils of this valley. This year was the latest I have ever seen it, being scarcely in flower by old May Day (May 13th). The adjustment of the calendar in 1752, by the sudden removal of nearly a fortnight from it, upset the festive association of mayflower and Maytide. The thick, sweet and sour scent of the may is a carnal odour and the festivities had much to do with the joys of the flesh, which is perhaps why the whole month was considered ill-omened for weddings, being more apt for lust than commitment. I doubt the enforced late flowering of the may had much to do with the riots against the calendar changes, the Reformation having long since done away with such pagan survivors in the Christian year. It's been said, condescendingly, that these riots were got up by credulous people who feared that they were being robbed of their full and allotted lifespans (not such an unreasonable fear in my view). The more prosaic explanation is that they were being robbed of twelve days' rent, and took justified umbrage at this sleight of hand on the part of the governing classes.

Eating disorders

Autumn 2012

Foxes, as is well known, kill for pleasure – at least, such is the appearance presented by a hen house after one or two have got in and had their extravagant way. Do they need to indulge in this slaughter? Some corpses which they don't devour right away they may bury for later consumption, and it is to be supposed that they cannot count and therefore don't know when to stop. All the same, once the killing is over they perhaps recognise that they've dispatched more than they can actually consume. It's likely that the feathery scrabbling and panic which their first attack induces brings on an irresistible murderous passion. Perhaps if the rest of the hens didn't flap so but just sat quiet they'd leave them alone.

So what's the badger's excuse? I don't mean its taking of chickens, which undoubtedly it does do (despite claims to the contrary), but its ravaging of horticultural crops. Until we started growing veg commercially I'd never thought of the badger as other than the shy, retiring and essentially benign creature spoken of in books and occasionally glimpsed with pleasure at either end of the day. My attitude changed when, after we'd been here a few years, we went out one morning to find that half of our quarter acre stand of early carrots had been obliterated overnight. The ground was littered with uprooted carrots and carrots still rooted but with their crowns ripped by sharp claws. There wasn't much sign that the badgers, for it was they, had actually eaten anything and I'm still not clear whether they dig roots because they are hungry or because they like the smell, or if they do it just for the hell of doing it. A little later they tried to have a go at the maincrop, but being forewarned we'd put up a couple of strands of electric wire and had the pleasure, as we lay in bed one night, of hearing agonised squeals retreating into the distance. It's not just carrots – parsnips too are grist for their mill, along with sweetcorn (of course), broad beans, potatoes, melons. One evening I checked a 30 or 40 yard bed of peas that we were going to pick for the boxes the next day. In the morning I couldn't find it, until I realised that the whole crop, pods and haulm, had been trampled flat into the surface of the soil overnight.

The badger population and with it the boldness of their marauding continued to grow and pretty soon a great part of the growing area, which occupied several small fields unhelpfully bisected by a public footpath, came to be skeined in electric fencing. When the MAFF mammal officer (sic) turned up to tell me that there was no way we'd get a license to control our badger population he observed that he'd never seen anyone take so much trouble to keep badgers out. But what else could we do? Rigging up a radio to a wet battery and playing them the BBC World Service all night had no effect.

The wood pigeon is now reckoned to be the commonest British bird and deer, in our case roe deer, flaunt themselves in a way that was unthinkable twenty years ago. Both are browsers by nature, but then I suppose that's true of most of creation, or that part of it that has large freedom of movement – taking the best of the day and while summer plenty lasts seeking out the most tender morsels and shoots without thought for the morrow. Be that as it may, the pigeon's wings and the deer's fleetness of foot lets them pick and choose. This ensures that the loss they cause to growers is out of proportion to what they actually eat, with the pigeon's dirty feet and incontinent habits just adding to the problem. The pigeon is of course the great despoiler of brassicas whereas my experience of deer is that, though they will ruin swedes by taking one bite out of each while going up the row, they care much more for lettuce, beets and umbellifers. When you add in the badger's activities in the root field and elsewhere that's all our common crops accounted for.

And what of the rabbit, which arrived here in the wake of the Norman conquest along with fallow deer and pheasants (another vandal in the vegetable beds)? Compared to the above they are a minor problem for us – it's usually a matter of a bite here and a nibble there – and as forty rabbits reputedly eat as much as a well-grown bullock I'm grateful for that. There have to be some advantages to growing vegetables on heavy ground and that rabbits thrive more on lighter soils is one of them. They too sometimes seem to get carried away by a destructive urge. I'm not suggesting that they spit out more than they bite off, but I've often noticed that they will dig up plants they don't care to eat, apparently out of devilment. Thus having planted out annual flowers in and around the crops I sometimes find these dug up and cast aside. For some reason they take against cornflowers in particular.

Lovers (as opposed to appreciators) of wild animals passionately maintain that badgers and the like have every right to eat, and so of course they do, but then we also must have some right to try to stop them eating our livelihoods. As the numbers of our predators increase and the awareness on the part of the public of the associated realities decreases the struggle doesn't get any easier. Once, leaning on a gate at dusk, I saw a badger out in the adjoining pasture field with its back to me and muzzle down in the ground, rooting for grubs. Protected species or not this was too good an opportunity to miss. Advancing swiftly and quietly across the darkening sward I'd just got up to it and was swinging my steel-toed boot back so as to administer a good kick to its backside when it turned, snarled and lumbered off. A second later I'd have made contact, a small but satisfying payback for all those years of pain and loss. On the other hand the possibility that in return it might have fastened its jaws around my ankle has subsequently crossed my mind.

"Pray you, tread softly —

Winter 2012

- that the blind mole may not hear a foot fall . . ." A mole has taken to throwing up heaps of earth just at the end of the saw horse where the logs drop off. Either I need to get a shovel to clear the landing space or the logs end up muckier than they already would be. What the mole is enjoying is a small carbon sink, not so much earth as a rich organic mix of decaying chainsaw chips and axe detritus which is alive with organisms up to and including worms.

The great bulk of wood that passes through the wood-yard and up our chimney has an aerial and more circuitous course to follow before it eventually gets back to ground level as fresh timber, but it is still essentially a local one. Firewood here is a product of hedges. There are overgrown corners and the stretch of bottom land where, advancing out from the brook, alders march and willows sprawl, but nothing you could call woodland. However the hedgebanks themselves make up for it. As best I can calculate there are about 2½ miles of hedge on our 30 acres. Taken all in all that would be close to a tenth of the farm area.

When we arrived here very little had been done in the way of hedge maintenance (or any other maintenance for that matter) for forty or fifty years, and nothing at all for the last ten. The entrance lane was so overgrown that it could only be negotiated on hands and knees. While a fair proportion has been cut and laid and brought back to order in the nearly thirty years since, there remains a lot of hedge that has now been untouched for at least seventy years and which groans with excess timber. Fortunately hazel predominates; of all trees the one most able to retain the semblance of a hedge even when neglected, through its constant self-coppicing. Otherwise you end up with a sparse line of standard trees which barely qualifies as a hedge at all. But even hazel makes a fair—sized clump when left to its own devices and wind-rock transmitted down to the stools, as well as the force of gravity exerted by leaning stems, will sooner or later cause the earth and stone banks on which the hedges sit to tumble down.

We are not short of firewood due to that legacy. But supposing one day all this length of hedge is restored and then maintained on a cutting cycle of 20 to 25 years, would this farm still produce enough for its needs? For most of its life it obviously has done. Until the coming of the railways and cheap coal there was no choice but to rely on what the trees of the locality could provide for fuel as well as for most of the materials of life. It seems that the idea of mediaeval England being thickly wooded, with settlements occurring as islands in a forest, has only romantic fiction to support it. When the Domesday survey was made in the 1080s Devon had little more woodland than it does now and the woods that there were throughout lowland England, were not free country but were owned and managed and coppiced for what they could produce. The peasantry had to make do with the leavings, but the farmer here had his hedges. I dare say that he felt he was well-off.

When at length we came to rebuild the old house, which unsurprisingly had weathered the years of neglect much less well than the hedges, all we could retain was its one short length of dressed granite walling. Inserted into the original cross-passage in the sixteenth century this had formed the back of the then new fireplace in the adjacent hall (that's 'hall' in the old sense, the chief room of the house and not the little entrance space that it has declined to since). Prior to its installation there had been a fire on the floor, the smoke finding its way out through the roof. In order to insert a second storey so as to provide a chamber above, the fire had to be moved to the side of the room and given a chimney. The fireplace was big – eight feet wide –but that doesn't mean that the household sat in front of a blazing fire. Only the gentry could afford such extravagance. Instead they moved the customary small fire, which gave heat for cooking and warmth to those sitting around it, to a snug space where six people could still companionably take the warmth of it from three sides. In cold weather the fireplace was for sitting in.

These days we expect to be able to take off a layer or two of clothing when we come indoors, and even then might want to keep our distance from the woodstove. Domestic warmth has perhaps done as much to increase life expectancy as developments in medical care – it is after all the old and sick that most feel the cold, which is a bitter thing to an ailing body. With better insulated houses and more efficient combustion we might make do with not much more firewood, but warmth alone is not enough, we need light and power too. Now, having exploited everything else, just firewood won't do. We're talking biomass – for burning in power stations, or even for turning into charcoal and burying in the ground in order to compensate for the damage caused by burning all the other things we'd have been better off not burning– and for this we have to look much further afield than the hedges and woods of England. But comfort, like charity, should begin at home. Not every hedge can be left to grow up, they interfere too much with crops and traffic for that, but if we harvested those we could and restored the active management of woodland we'd be getting back to doing what our forebears did out of necessity, which is to make the best use of the resources to hand. It's not an answer, but at least in doing so we'd be restoring the battered fabric of the countryside – its value as habitat, its useful pleasures and its functionality.

NATURE NOTE 15

Almost a vegetable

Spring 2013

The one pleasant surprise among my small and almost permanently saturated area of 2012 field vegetables was a few handsomely large swedes. A few large swedes is almost a contradiction, because this is a crop known for its liking of being in a crowd of its own kind, in a field of its own where the wind is free to roam. We used to grow an acre or two for the wholesale market, but with boxes swedes became the one thing we did habitually buy-in. I'm only growing them now because I'm more of a gardener these days and can please myself as to what's worth growing, and because the ones available to buy lack sweetness and are meagre and tough.

Maybe you can still get a decent swede in Scotland. I hope so. Nearly forty years ago I worked on a farm in the north-east Highlands where we grew more than thirty acres (one field) of them, about 10% of the farm area. They weren't called swedes of course; usually neeps, sometimes swedds. Through the winter they fed both beasts and man, chopped in the first case, generally mashed in the second. With the addition of a bit of rolled barley they fattened finishing cattle, with hay they sustained the stores, while they kept the sheep going pretty much on their own. As for man – there wasn't a great deal else available in the vegetable line through the winter, and these were good swedes.

This was my first experience of row crops – little did I know the life to come. Mildew not being an issue up there they were sown in early May on wide ridges and grew mightily, an average individual being 4 or 5 pounds in weight. Once or twice in early summer men came by in hope of work singling and hoeing, an echo of the time when agriculture depended on such itinerant labour. They were too late – a Webb's precision drill had done away with the need for singling, and as for hoeing . . . an applicator on the drill dispensed pre-emergent herbicide, a flowable powder (I believe it's called trifluralin) which somehow spared the emerging neep but gassed everything else. I never picked up a hoe until I became an organic grower.

Weeds did grow though, as they will, especially on the sides of the ridges, and this gave me my first experience of mechanical weed control. We used a scarifier, elsewhere known as a side-hoe, to cut them away. This consisted of a pair of angled discs for each of two ridges, each pair being shiftable laterally by means of long levers that could be reached from the tractor seat. This and the drilling was when the grey Fergy had its few days of glory on an otherwise more highly mechanised farm, and pleasant it was to putter up and down the drills, with (on the upward bout) the great waste of moor and mountain rising one farm beyond us to stretch away in its emptiness to the furthest coast and then – on the downhill – a neat and peopled country held in view between two arms of the eastern sea. An additional pleasure was the presence of several pairs of oystercatchers, which in the north have over two or three centuries extended their nesting zone far inland. The nest itself – an inconspicuous scrape marked by nothing more than a few small pebbles and a straw or two – might as well be on the beach. But there is nothing inconspicuous about the bird, one of our largest waders. Its orange-red beak and eye and long pinkish legs are complemented by a smart piebald plumage, black above with white breast and underparts, and white again revealed on the wings in flight. It has too a shrill and insistent piping call which it is not shy about uttering. The whole ensemble gives it a bold and confident air, wary but not cowed. They'd sit tight on their eggs until you'd worked quite close. I'm sure it's true that you could move the eggs out of the way and replace them afterwards, but I preferred to arrange things so that I could let the nest and eggs pass undisturbed under the centre line of the tractor.

Back on a mixed farm in the Westcountry we grew a few acres of swedes most years, more as an afterthought than as the central plank of winter feeding. The variety was Marian, then ubiquitous – presumably on account of its resistance to clubroot. It certainly had nothing to recommend it in the kitchen, and – tilled about Midsummer as is normal down south – made an unimpressive crop compared to what I had known in the Highlands.

Later (on our own land) we discovered that it was such little starvelings, as I thought of them, that the market wanted but at least we found a variety that cooked and tasted well. This was Devon Champion, bred and maintained by the venerable Edwin Tucker and Sons Ltd of Ashburton. Even so it was a relief to give up wholesale swede growing. The root fly was bad enough and we would certainly have been defeated by the rising tide of flea beetle through the 1990s. Devon Champion, once the king of the Devon swede market, is available no longer. It doesn't process, and that's the end of it. This season's swedes are Ruby. They're not bad, better than expected, and what's more – I never saw a flea beetle in 2012.

I remember a student meal, when there was little in the larder, that we christened Massed Swede. It left us yearning for something else. The swede is good stock-feed, but a poor sort of vegetable. Some vegetables will make you most of a meal, others are at least adaptable and can be deployed in a variety of ways. But the swede – you can mash it, or you can put it in a stew. Either way it wants a bit of meat to set it off. As a crop it turned out to be a sort of bridge for me between agriculture and horticulture, but I think it's best left on the farmer's side of the line – almost but not quite a vegetable.

Bird business

Summer 2013

We were surprised to see eight swallows on the 14th April, early for here by a fortnight. The bitterness had at last gone out of the weather, but it was still cool, windy and damp. These must have been the first swallows for many generations to witness snowdrops still, just, in flower. Impossible to imagine what sustenance they could find aloft – apart from two widely separated and, I imagine, somewhat desperate bumble bees I'd seen no insects on the wing since some winter gnats in about December. It was only the previous day that I spotted a blackthorn bush just chinking open its white buds. By the season these swallows were more than a month ahead of themselves – what on earth (or rather, above it) were they going to eat? But then the fifteenth dawned like a miracle into warm sunshine and Spring, pent up and urgent, came tumbling in. At last the celandine – long caught between the conflicting currents of lengthening days and bitter temperatures, their reluctant flowers burnt by the icy winds – could do what they are put on earth to do, which is to reflect the spring sunshine in bright sheets of yellow-gold. Now earth and air were charged with life. Suddenly insects were everywhere. That very day I saw both tortoiseshell and peacock butterflies. This was like springtime in the Highlands, a glorious effusion. More often the Devon spring starts in a hesitant way not long after Christmas and sputters on until eventually overtaken by summer some months later.

The change had come a year almost to the day since the equally dramatic shift in the weather pattern last April. Not that it's stayed balmy over the four weeks since – there was a brief blackthorn winter when those flowers fully opened, and it's been more windy than not – but a corner was turned. Now those of us on heavier ground need a drought. It was striking and much remarked on locally that every little bit of rain we had during February and March immediately turned the surface back to a state of soddenness, despite there having been plenty of time since the floods for the water table to subside. Eventually I realised that the constant rainfall of last summer had prevented the cycle of wetting and drying and the consequent fissuring which is essential for soil structuring on this land. Without it water can hardly drain away.

So now our pair of swallows are domiciled once more in the loft of the packhouse, where they'll make a devil of a mess all over the timber and assorted horticultural detritus stored there, while delighting us with their liquid flight, pleasant twittering and the doings of family life. No sign of them sitting so far and despite its tumultuous arrival Spring has some way to go. A fortnight after May Day and the may flowers are still wrapped in their darling buds. The celandine though have become a pale memory beneath the jostling uplift of succeeding herbage.

I say "our swallows" but of course they are not ours. It may be this unconscious possessiveness that as much as anything else sets civilised man apart from the animals. At its best it leads to curiosity and a desire to name, classify and understand. At its worst it says 'we are the crown of creation and all the world is our domain'. Somewhere between it leads to ignorant anthropomorphism and a desire to make the wild tame. As among the wild animals it is only birds, some of them, which will let us engage with them on more than a fleeting basis it is no great wonder that the RSPB is famously the largest conservation society in Europe. I remember sitting through a dull address by a RSPB spokesman at one of the Cirencester conferences, a manifestation of long drawn-out diplomacy as the organic movement attempted to get it (and thus its substantial influence) to come out in favour of organic farming. It wouldn't commit itself then but I'm glad to see it does so now. One of its catalogues arrived in the post a while back. The cover says "Save nature while you shop", as dubious as the claim (made inside) that its toiletries "are good for the environment" (as if the environment is in need of toiletries), but much of what it's selling is well-designed and decent and some of it no doubt useful to all degrees of bird-watchers. I wonder though about this commercial edifice built on bird life and whether it doesn't take on a life of its own, the birds becoming figures in a consumer landscape. There are questions too about whether we should be interfering in their lives at all. The catalogue has a good bit to say about the importance of cleaning your bird-table, but is too coy to mention the trichomonosis parasite that has devastated the chaffinch and greenfinch populations in the last decade. The fact is that bringing a constant succession of birds to countless bird-feeding stations is a long way from how the birds themselves would organise their lives without this human inducement to congregate. It shouldn't really be a surprise if the consequences for them can be fatal.

I'm pretty sure we can't get by in a world which compartmentalises "nature" into, for example, Environmental Stewardship options or RSPB reserves while the rest of the land becomes an agricultural desert. Part-way domesticating songbirds in suburban gardens may be a reasonable response to such a state of affairs, but if you feed them please wash the feeders every week. The Jet 5 that you may use for disinfecting horticultural propagating gear would probably do it, or the RSPB will sell you a scraper, a brush and some disinfectant (pg. 51 of the catalogue).

NATURE NOTE 17

The swallows' summer

Autumn 2013

Two or three days before they might have flown the first brood of swallows disappeared, leaving just some sad feathers on the floor of the loft. I blamed the cat and then myself – for not having shut the trapdoor at the top of the stairs and removing the stack of produce crates that provided it with an alternative upward route. The parents tried again and it was during the course of this second attempt that I realised that the cat, though guilty no doubt of many atrocities against birdlife, was probably not to blame. (In any case, viewed dispassionately, it was hard to see how it could have got at the young ones while still in the nest, fixed to a metal purlin at about head-height from the floor). The loft covers the whole of the two outside bays of the shed but only the back of the centre bay. There is thus an apron at the front into which tractor and link box or van can be driven for loading and unloading. The loft opens directly onto this space, although in a half-realised spirit of health and safety the edge is guarded by wooden rails on two of the three sides. At the back of the concrete apron, against the block wall of the veg store behind it, is a large wooden trunk which my father had made in Egypt at the end of the war so as to bring his booty home. It now holds our extensive collection of gumboots and makes a useful seat on which to rest weary limbs and note the comings and goings above.

I was in and about the shed one day after the swallows had hatched their second brood and been feeding them for a while, when I heard a sudden commotion of furiously anxious parent birds. Looking up I saw a large shape close to the nest which then came to rest on the rails at the edge of the loft and resolved itself into the menacing figure of a hawk, briefly fixing me with a downward shaft of its steely eyes before flying off. The chicks survived that close encounter, but a week or so later the nest was empty once more, the only trace being a scatter of stubby quills. A sparrowhawk seems the most likely actor for this kind of deed, but peregrines nest only a few hundred yards away on the craggy face of a disused quarry and perhaps it was one of those. In the confusion I didn't get much of a look, and anyway am not much of an ornithologist.

Brave birds, or merely programmed to do what they must do, it matters not. We were now so emotionally bound up with the trials of their breeding summer that the swallows' third attempt had us on tenterhooks, checking their progress with upward glances at every opportunity, willing on the parents as they quartered the air for flying insects and zoomed in to the nest and out again. It was a release and a matter for great rejoicing when, on September 12th, the four youngsters were off the nest – still under the roof, but stretching their limbs. On the 13th six swallows were flying aloft in calm winds and light rain and then the following day, a breezy Saturday, we watched the fledglings enjoying the sunshine in short flights from the phone lines. Since then it's been uncomfortably stormy and today, the 17th, the birds are still occasionally in the shed and the young still being fed by their parents. But it feels like we've done our bit, which is to say our passive hospitality has gone a little way towards keeping the breed going. Swallows and martins both are scarce this year. Normally on a sunny day at this season the telephone wires along the road into the village are lined with young hirundines like so many ripe fruits, but there are few now to be seen.

Unlike the swallows which arrived early, the spotted flycatchers came back to their ancestral nest in the other shed so late that I'd despaired of them. But there they were on June 1st and then lost no time in laying, incubating and raising four fine fledglings – all within six inches of my upstretched fingers. These flew in the first week of July. I glimpsed the little family for a few days more, as they moved away down the hedge. But the parents never returned for another brood, as they have done every other year. There have been plenty of bumble bees this summer, but other insects – aphids, hover flies and ladybirds among them – have been in short supply. Houseflies too, and though hardly regretting their absence you have to wonder as to the cause of it in a summer of such heat. The swallows managed to feed their three broods, but had the air above the farm mostly to themselves and faced little competition.

Across from my seat on the wooden chest a gap in the hedge opposite lets onto the old back orchard of the house. A tall gangling apple tree, one of three survivors, stands just through it. Its trunk has been quite hollow ever since we moved here. When we lived in caravans it was within reach of both ours and our daughter's (she got her own when she went to secondary school). I remember us looking out the window from our bed one stormy morning, twenty-five years ago and more, and watching it dance frenetically in the dim light of dawn as the roaring wind tugged it this way and that. Jan said to me "it's quite safe, isn't it?" to which I reassuringly replied "I don't know Jan, it's like Russian roulette". But it's still there, a Sweet Alford, and still bears masses of pink-flushed yellowy apples – for cider, insipid, like a Golden Delicious but without the taste. Beyond that and beyond the further hedge a scrubby wood of ash and sycamore can be seen rising steeply on the other side of our little valley. Against that backdrop, just yesterday, I witnessed a small group of swallows defiantly and vociferously mobbing a peregrine. It flew off, swallowed from sight against the dark of the trees.

The night

Winter 2013

In his 'Wild Life in a Southern County' Richard Jefferies relates that the shepherds on the downs speak of occasional nights of such impenetrable blackness that sheep are panicked into breaking down the hurdles which confine them. It is improbable now, getting on for a century and a half since he wrote, that many places in any southern county could experience nights of that intensity, and certainly not Jefferies' native farm (which you pass by and perhaps over as you speed along the M4 by junction 15, the Swindon turning) nor any of the surrounding country whose detail and essence he transmuted into that wonderful book.

The sad fact is that the night is not as it was. The glare of artificial illumination seeps out from town and suburb, from business park and motorway junction, dissolving the darkness and staining the atmosphere in an excess of casual energy. It is often bemoaned that a large proportion of the population have no idea that milk comes from cows or that potatoes grow in the soil, symptoms of humankind's estrangement from the land. A much bigger proportion know nothing of darkness – I mean the living darkness of the night, not what you get when you retreat behind thick curtains and turn the lights off. It's true that humans are adapted for daylight and have little use of the dark for other than sleep, but still this is an estrangement too, because it is the night that reveals the cosmos to us.

We are lucky here I suppose. In this hilly country the relatively feeble streetlights of the village half a mile away are hidden by rising ground and the night view has no more than a few points of light. There is however a moderate sized city about eight miles to the north-east which grows and gets brighter year by year. It lies behind and well below the high ground that marches between its valley and ours, and in daytime there is no sign of its existence. But by dusk it makes its presence felt. In clear weather only the brightest stars are visible until well above the north-eastern skyline, so that at this season the rising Capella is for some time deprived of her accompanying kids (three fainter stars in an isosceles triangle). Its effect is greatest when the sky is overcast – then the city's light is held down and reflected off the bottom of the cloud cover and thus spread out over a wide extent of country, not just its own valley but this one too. At such times it makes a notable difference to visibility on the ground.

Sometimes though the cloud base is on or below the intervening hills. Then the city's presence is altogether blotted out and, free of its influence, the night waxes blackly. Navigation becomes possible (in this land of tall hedgebanks) only by looking upward to the faint delineation of black vegetation against the other blackness of the air. On such nights the fifteen minutes or so that it takes the human eye to adjust to darkness makes less of a difference to visibility than when there is some shred of light for the eye to work on. It is interesting to speculate as to how animals experience dark and light. Cattle, which are mostly diurnal in their activity, will gaze into the beam of a powerful torch without blinking, as if it's not there – but this may be a symptom of their general vacuity. Cats, which are nocturnal as much as they are diurnal, can sometimes be induced to chase a torch beam (as kittens) but otherwise seem not to differentiate light and darkness. Badgers, though I believe badger-watchers employ red filtered torches, in my experience take no notice of ordinary torchlight. Birds, most of them, are like us and don't much care for the darkness – but many of them make their migratory flights by night.

So the birds are at roost and we're keeping the night at bay with artificial light, but if we never see the night because we stay indoors or live in a place where darkness has vanished we miss out on the glorious phenomena that it sometimes shows. I thought I'd have to go to Canada to see Aurora Borealis, the Northern Lights, and so I did – as a faintly shimmering greenish curtain – from an enchanted island at the mouth of the Moose River in northern Ontario. But then we saw them here way down in the south of England on various occasions in the late 1980s and early 90s, usually a rather lurid pink with slowly moving shafts of light, as from a car's headlights, playing on the curtain. It's down to solar activity and then on being outdoors to see them, in our case washing leeks and so forth on the forecourt of our packhouse while decent folk were indoors in front of the television.

Another phenomenon I'm pleased to have seen on a few occasions is the moonbow, not a halo round the moon but a rainbow of moonlight. My veracity has been challenged on this matter but they do occur; like the daytime version but white or silvery grey and even less substantial. The moon has to be full or nearly so and not far above the horizon, and of course there has to be rain.

Depending on its intensity light pollution may do for these phenomena, as it does for comets and shooting stars. It certainly does for the full glory of the stars. Whereas every child could once have recognised Ursa Major (the Plough, the Dipper, Charles's Wain, call it what you will) it now goes about its business of pointing the way to the pole star to the ignorance of most. What chance is there that the most obvious constellation of the circumpolar stars, those we see all through the year, should unfailingly indicate the signpost of the north? There's a part-time astrophysicist, otherwise a musician, who has a lot to say about badgers. I've got a bit to say about badgers too, but nothing to say about astrophysics. I don't know about an ever expanding universe, which to me sounds like an unending nightmare. I prefer to see the clear night sky as a great tree, its branches hung with lights. Their arrangement does not seem altogether random. I think there may be some other kind of truth here, but what form it takes I wouldn't care to say.

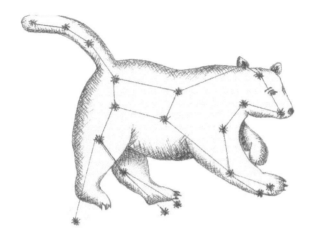

NATURE NOTE 19

Land and water

Spring 2014

The rain gauge has been working hard lately, so much so that at times it's felt as if the world itself was washing away. It recorded thirty inches between mid-December and mid-February (our annual average is 45 inches) and apart from December not having been the wettest ever (12.2 inches in 2000) every other record has probably been broken. Flooding on The Levels prompted memories of an advisory visit I made some years ago to land beside one of those cut and embanked Somerset rivers. The clients were loud in their complaints against water voles – there were too many of them, they undermined the river banks and wrought havoc in gardens and allotments, burrowing and eating root crops. To my surprise they felt as strongly against water voles as some of us do against badgers – indeed the damage they ascribed to them was of much the same order as the destruction we hold our badgers liable for. The difference is that while badger numbers have increased markedly (I recently saw a fit and active specimen on the forecourt of our village post office at 12 noon) Ratty's progress nationally is all the other way. The non-dredging of the rivers that we've heard so much about may not just be down to the Environment Agency wanting to save money, no doubt preserving voles gets a mention in the policy document too.

No water voles here, more's the pity! The watercourses are of the wrong sort, being stony, rapid and largely free of weed. Years ago I knew well an area of flat, agriculturally ill-considered land occupying a shrinking space between heathland to the north and suburbia to the south (it was the suburb, not the heath, that was increasing). It's gone now, but in its long day the several roughish pasture fields watered by two streams that joined to make a small river were a haven for wildlife and in particular for aquatic life. One of the streams rose on the chalk and was especially well charged with nutrients, decorated with an exciting flora and well-populated with animal life – including large numbers of voles. Walking along the bank the plop of water rats diving to hide themselves beneath the surface was a pleasantly frequent sound. In fact very little patience was needed to catch sight of one sitting at the entrance to its burrow vigorously chewing on some vegetable matter, or cruising around, head up, in the easy current. It was here too that I once saw a grass snake swimming, or rather gliding, over the barely disturbed water. The grin (I'll call it that) of pure pleasure on its face has remained with me ever since.

Then one season some authority or another took it into its head to get the stream dredged, for no good reason that I could see then or can think of now. Its crinkly banks were dug up into a uniformly sloping profile, its bed shaved level and all the life of it mangled and destroyed. Disaster! And yet within a couple of years all the life was back – the voles, the snakes, the gudgeon and bullheads, the loosestrife and the bur-reed – so quickly can such habitats recover. Water spreads the seeds of new life along its course and carries also the silt and nutritive material that lets them flourish. Left to its own devices it soon reclaims its own.

With so much water currently on the ground and bursting out of it, and our several tiny streams charged to the brim and proudly noisy, it's struck me how much of the original land clearance on this farm and on many others was to do with controlling water through drainage. This was perhaps a bigger task than the clearing of woodland which one thinks of first in this connection. Much of that, given time, could be achieved by grazing animals and pigs, whereas channelling the water was entirely reliant on (to use a Westcountry phrase) louster work, that is on physical toil. Without that foundational work the many springs at the base of the steep escarpment above us would seep over much of the more gentle slope that is the better part of this farm, and which otherwise might bear little but swampy alders and rushes. Instead they were captured and led down along the field edges at the base of the hedgebanks, thus as well as drainage also providing water for stock and, later in the history of the place, the means to irrigate winter pastures (which flushed grass growth by raising soil temperature) through an intricate system of shallow leats.

After a short life of at most a few hundred yards these little streams discharge into the greater one, the Rookery Brook, which forms our bottom boundary and also the boundary of the parish. Rather belying its pastoral-sounding name the brook has fallen 600 feet in two miles by the time it has got to us, emerging out of a steep and narrow cleave and still tumbling noisily over the boulders that litter its course. Well confined in its valley, the brook has so far had little problem in conveying the great excess of water to the river half a mile beyond with just some minor spillages on either side.

Somerset's chief river, the Parrett, falls 11½ feet in the 11½ miles between Langport and Bridgwater. Draining the flat and barely above sea level expanse of the Somerset Levels was obviously a challenge of a different order and quite beyond the means of individual peasant cultivators. It took the corporate presence of Glastonbury Abbey to get it under way in the thirteenth century and the process has continued at various periods since. Lord Smith and his staff are the successors to the abbot and monks of old.

The winter of 2013/2014 was, according to the Met Office, the wettest for 250 years. In particular, large areas of the Somerset Levels were under water for weeks and even months on end. The Environment Agency, responsible for its drainage, came under a lot of criticism. Lord Chris Smith was its chair from 2008 to 2014.

Life chances

Summer 2014

Willow down drifts by outside the window on a now gentle breeze. More strength in the sunlight today, less of that nagging wind from the northwest, and so the sallows release their ripened seed into this hopeful warmth. It goes by in gentle swarms, specks of questing life borne on countless little assemblages of fluff (the tiny seed needs none of the cunning architecture that floats the dandelion). These gather on the spiders' webs in the corners of the casements, revealed now that the afternoon sun begins to light up the back of the house, and anywhere in its path the spiders must put up with their labours being undone by the mass of down that coats their sticky threads. The quantity released is so great that the bare ground of the footpath beneath the hedge takes on a whitish cast, and here and there – where it gathers against a fallen twig or other slight obstruction – it forms little drifts, like so much fairy snow. There is of course a reason for this fecundity, which must require a considerable output of energy on the part of the tree. The seed germinates readily, but only if it meets moisture during the seven days that it retains viability – some that has come to rest in a tub by the back door (which for the moment contains just a few inches of rainwater) is already transformed into floating pairs of little seedling leaves. But the chance of any one seedling surviving its first summer against the perils of desiccation or of being shaded out, or grazed, mown or trodden out of existence, is minutely small. Not that the sallow is dependent on seed for its continuance as, bramble-like, its limbs bend down and root again where they touch the ground. In this way it welcomes the storms that half break its branches and rejoices especially in a heavy fall of wet snow which bows it to the ground. Left unchecked it is capable of advancing steadily across dampish ground without any need for new seedlings.

Spring fecundity is not confined to the plant world. A backwater of the brook teems with tadpoles, but if just half a dozen of them return when they've reached sexual maturity in three years' time equilibrium will be preserved. Less well-known but welcome to me now as a token of spring is the oil beetle. It's true that it is a poor substitute for the vanished cuckoo, but the sight of this portly, purple-black, inch and a quarter-long behemoth determinedly pushing its way over and through the awakened herbage is reassurance that at least its part of the web of life remains intact. Emerging from its subterranean chrysalis sometimes as early as March the female oil beetle ("oil" referring to the defensively unpleasant fluid that this beetle secretes to warn off potential predators) must first find a mate to fertilise her eggs. Having done so, and afterwards choosing some flowery place which offers nectar to flying insects, she lays them in batches of up to 4000 in little scrapes dug in the ground. This magnitude of egg-laying signals the convoluted game of chance that the resulting larvae are born to. The eggs hatch into active little creatures about one millimetre long, in which form they were originally known as triungulins, or bee lice, no connection between them and the oil beetle being recognised.

Each is equipped with six grasping legs and a glue-secreting gland at its nether end, and by these means they satisfy their first instinct – to climb up into the petals of a flower. Their second is to await the transportation which is their slim chance of a ticket to eventual adulthood. They will attach themselves to just about anything that happens by – fly, wasp, bee, spider, or even a blown straw – but only those solitary bees (*Anthophora species*) which provision cells with honey before laying an egg in each one will do, any other mount leading only to oblivion. If they have the right bee but the wrong sex it seems they can make good their mistake by transferring host at the moment of the bees' mating. Once lodged on the back of the female the louse (of which there may be several per bee) does its best to cling on through all her otherwise solitary nest-making preparations until at the appropriate moment one and one only hurries to the abdomen and attaches itself to the egg as it is laid into the honey-filled cell.

Now at last the little creature gets a meal, biting through the egg shell and devouring the contents. Using the empty shell as a raft against drowning in the honey it then settles down for a protracted moult from which it emerges in an entirely different form. Never again will it look the least like a triungulin! Now it is fat and squat, its belly a deep keel capable of supporting its smooth body as it floats on the surface with its jaws, themselves transformed into paddles, sweeping the honey into its mouth. This is not the end of it – from beginning to end five separate metamorphoses take place, the last being the springtime emergence of the adult from the underground winter pupa.

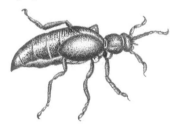

And there it is, this ugly and ungainly beetle which loves the spring sunshine and seeks out its flowers, lumbering around with no attempt at concealment. The oil beetle's success, unwelcome as it is from the bee's viewpoint, is also testament to the survival of the bee – the first not being able to do without the second. An extraordinary story in itself, a vignette of nature's curious ways, you may think it extraordinary too that anyone ever came to work it out. It would be interesting to know how evolutionary theory accommodates this tortuous tale.

Spiders, from Mars?

Autumn 2014

Like it or not any vegetable grower has a relationship with arthropods – what are commonly known as insects but which include non-insects such as woodlice, centipedes and spiders. Some are actively involved in our professional life either as pests or, being predatory on those pests or pollinators of our fruiting crops, as allies. The majority have no obviously direct bearing on our growing activities but may be appreciated just for what they are and for the interest and sometimes gaiety that they bring to our working lives. Probably most of us dispose of slugs and squash destructive caterpillars without much of a qualm. In our early days here I dispatched hundreds of wireworm – it made no discernible difference to the total but made me feel that I was at least marginally redressing the balance. On the whole I rather cherish the multitude of small life-forms that we share our land with, rescuing little creatures that find themselves trapped in buckets of water and ushering butterflies (other than cabbage whites, which I do my best to kill) out of polytunnels. The other day I grasped a large ichneumon – wasp-waisted flying insects that parasitise caterpillars – which had come to the light above the kitchen table and then got caught up in Jan's jumper. My idea was to put it out the window, but the searing pain experienced when it thrust its ovipositor into my finger made me let it go. After that I cared little for its fate and it will be the last time I lend a hand to an ichneumon. But the principle stands.

There is though a spider which really gets my goat. It does us no actual harm beyond insistently sharing our domestic space and indeed it must reduce the number of cohabiting flies and the like. As spiders go it is clean in its habits (having little in the way of a web), nor is it in any way scary or threatening. Between April and October it hangs around the house in great and ever-replenished numbers, mostly between wall and ceiling but also in other spaces – anywhere that offers an internal right-angle will do. That's one thing, but the real cause of my irritation is that despite having shared our house with it since the day we moved in and the ample opportunity that I've had of observing its ways I am no wiser now as to the secret of its success than I was fifteen years ago. It just sits there, motionless and inscrutable.

Of course it's not really from Mars (I understand it to be French in origin) but for all that I can get a handle on it, it might as well be. Its name is *Pholcus phalangoides*, sometimes called the Daddy Long-legs Spider, and it's said to now be outcompeting all other house-dwelling species. When standing on a flat surface with its legs neatly arranged around itself it looks elegant enough, but for the most part what you see is an ill-defined shape surrounded by an untidy gathering of spindly legs, the whole suspended a little way off the solid surface by a thread so fine as to be invisible. In the dim recesses of a shed (which I think of as their natural habitat) they tend to be a ghostly grey, almost colourless, but in the stronger light of the house they vary from a pale to dark brown. Either way they lack solidity – there is something spectral about them. The head and body of a fully grown specimen is about a third of an inch long, while its legs span two inches or more, but they start out as little blurry pinheads. Strangely there is always a range of sizes present, from tiny to large. It's odd too that a creature which spends its life indoors is apparently highly weather sensitive. If you clear them out of a room and the north wind blows for a week then it will be more than a week before the population regains its former state, but should the weather be hot this is achieved in two or three days. Again there is the full array of sizes. How do they do this? And where do they go in winter, and why? Perhaps the odd one remains lurking behind a cupboard or under the bed, but otherwise they just vanish.

You don't expect spiders to socialise, it's not something they're known for. But when there are so many of one species, sometimes stationed less than a foot apart, then you might expect some sort of interaction. I once spent (wasted?) a quarter of an hour watching two apparently making for each other, but they passed like ships in the night. Of course they breed, you can see big females carrying their parcels of eggs, but they must do it when the lights are out. As for eating – occasionally they do catch small flying insects, but they don't appear to be much good at it. The webs are tiny in extent and very fine, only becoming visible once they've gathered a bit of dust. Generally the little gnat-like flies that are attracted to light on summer nights evade them with ease. Capable of a loping run on a flat surface the spiders' movements are otherwise faltering, but mostly they don't move much anyway. Several days in one spot is not unusual. They do however have a singular method of defence, which is to spin gyroscopically on the end of their short thread like a whirling Dervish, becoming a blur of movement. If you want to grab one (and sometimes when I'm feeling charitable I chuck them out of the window in preference to leaving them for something more drastic) you must do it quickly.

We now seldom see the old-fashioned bristly house spider, the Tegenaria, but there was one until recently that lived at the back of the w.c. and would rush out to challenge my big toe when I went for a pee at some dead hour of the night. How the effete and attenuated Pholcus can outcompete the aggressively muscular Tegenaria is a mystery so far as I'm concerned, but I know which of the two I prefer.

Where are the birds?

Winter 2014

Usually by now cold weather, somewhere else if not here, has caused the arrival of an advance guard of migratory thrush species – blackbirds, thrushes themselves and most notably, because we don't see them at other times, fieldfares. Even if they know we have hardly an apple this year there are holly berries aplenty, more than I ever saw before. It will be nice if there are berries still at Christmas, but all the same there is something not right, the season seems somehow constipated. Where are the birds?

There is always a lull in bird life once the young are reared and the parents go into their late summer moult, so while the swallows are still flying above and feeding their last brood most of the other familiar species tend to be skulking somewhere out of sight. The first sign of new activity, the beginning of a new cycle of claiming territory, of pairing and breeding for the stay-at-home birds and a harbinger of the arrival of winter visitors, is the singing of robins in late September. It's a song of heart-tugging sweetness that in its clarity always summons thought of winter air and frosts to come. There was little of autumn about this year's September but the robins sang all the same. Since then – almost complete silence and hardly a bird to be seen, little more than the ever-present dunnocks on the ground, the occasional explosive trill of a wren (56 notes in 5.2 seconds!), ravens grunting their way across the sky and, always, wood pigeons. I'm actually thankful that these are now silent too – their tedious refrain of 'well f- – - you farmer' gets on my nerves while reminding me that their current destructive ubiquity is down to the unbalanced mess that agriculture is making of the countryside.

I've never picked courgettes so late as November 5th before. It's as well I did pick them then because there was a frost that night and the plants that had been so lively and proud one day were glassy and blackening the next, but before the morning was out it was back to Atlantic rain. This first lick of winter, slight as it was, was enough to induce faint signs of the autumnal stir-up – normally as substantial in its way as the coming of spring. Looking out on the back yard there was a cock bullfinch showing his finery off against a clump of polypody ferns on top of the old stone wall and behind him a spotted woodpecker, equally arresting in its plumage, tested the leaning poles of firewood in the back yard for grubs. Out front four song thrushes were investigating the possibilities offered by our small lawn while three cock blackbirds chased each other through the lichened branches of the ancient house-high plum tree beside it. A nuthatch worked up and down the trunk of the plum, intent and quick, and in the old orchard beyond a couple of jays flew circumspectly between its sparse apple trees. Later I saw a little flock of chaffinches searching out seeds from the weedy hardcore in front of the shed. Nothing unusual in any of that – these are all birds we expect to see at any time. But since then it has been as quiet as before.

What we do have is rain. I spend six months of the year feeling that I might drown in greenery, and the other six half-expecting the farm to subside altogether into the swamp. A friend who grew up in our neighbouring village tells me that as a boy he never experienced mud – that's the difference between life at the top of the scarp on the sands and here at its base in the silts. It's been tough for the worms lately. Through a design fault there is barely a step between the outside world and our front door. As the ground fills with water so worms come up and, seeking to avoid death by drowning, wriggle their way into the porch. Many take refuge in damp shelter under the gumboots, but several carry on under the door and end up as dried-out specimens stuck to the slate floor of the cross-passage. That's if I haven't collected them up and chucked them out, but it's not easy picking up a worm and they obviously don't like it. I believe (whether correctly or not I don't know, but this is what I've been told) that the touch of a warm bodied human burns them.

Bird life, by any reckoning, is fading. Country people shouldn't need surveys to tell them that, but then again the ones working the land only ever have to get down from their hermetically sealed tractor cabs when they need a to take a leak. Our local invaluable contractor can drive just once around a field and the computer has mapped it. Then, should he want to, the wonders of GPS allow him to set his course according to the screen rather the land. So what birds are there in that view, and should we wonder if the agricultural landscape becomes a map of functionality, shorn of its life?

The RSPB has sent me one of their frequent catalogues (I don't know why; I've never bought anything). What can we look for in that direction? Bird food certainly, and Christmas cards. It's said that you only ever see two robins co-existing peaceably on a Christmas card, and they haven't altogether avoided this pitfall, the maximum number on one card being nine. What really caught my eye was 'Forest Friends – a badger and a hedgehog frolic in the snow!'. Good job they've kept them to separate cards (five of each) otherwise, before they reach their destination, there'd be ten inside-out hedgehog skins and ten slightly chubbier badgers. Wildlife would be better served if treated with honesty and respect for its true nature.

Frogs on the other hand are doing well, around here anyway. I've never seen so many little ones in previous years as I have this. The other night, in pouring rain, there was one in the cat's bowl just out from the porch. It sat there, up to its neck in water, with all the concentrated self-possession of a toddler on a potty. I couldn't help but laugh.

NATURE NOTE 23

Light

Spring 2015

By Christmas Day the sunset (given equality of weather conditions) is already just perceptibly later than it had been a fortnight before, whereas through some quirk of calendar and/or heavenly bodies the mornings continue to darken until into the new year. There is nothing like the unfriendly murk of an early January morning, especially when compounded in the slithery mud of a winter vegetable field. It always made it a particular struggle to get back to work after the unaccustomed break that came with Christmas. But at the other end of the day the sunset's daily retreat of a minute or two is cause for hopeful joy. It is not just that it stays light later but that the light takes on a different quality. Whereas before the solstice daylight seems to stiffen and congeal into darkness as the gloom comes on apace, by Twelfth Night you can already feel the light being stretched and attenuated, as if reluctant to follow the sun over the western skyline. And so the day lingers a little and with that comes, out of both instinct and memory, the knowledge and hope of the warmth and awakening to come.

It is perhaps the changing quality of the light through the year (especially noticeable at dawn or dusk) as much as the seasonal weather that is responsible for a trick of recollection that I have been conscious of in my time on the land. This is that when thinking back to work done on the farm and the conditions and circumstances of its doing it is easier to summon up an image of twelve months ago than of nine, six or maybe even three months past. Perhaps this is just a way of saying that you can almost bring yourself to disbelieve in the possibility of mud when the ground is dried out and dusty, and vice versa, but it is certainly true that light is of itself one of the organising principles of animal life – witness the fact that it is diminishing day-length that induces ovulation in ewes and extending daylight that brings hens into lay. The skylark rises to greet the first rays of the sun before they've yet touched the ground beneath, and humans too are not unmoved by the rhythm of light.

In the world of green plants light is the one organising principle, and the source of life itself. It is there in upward growth (or if not upward at least light capturing – as for instance in the spreading and smothering leaf rosette of the dandelion) and the apical dominance that comes with it – to him who hath shall be given (and devil take the hindmost). Of course, the process is modified by warmth and moisture or the lack of them but the impetus for a plant to grow lies in its reaching for light, while the lengthening or shortening of the day is often crucial to its development of flowers and fruit. This is how we end up with the idiocy of French beans needing to be floodlit when grown in the tropics and the equally artificial illumination of AYR (all year round) chrysanthemums.

There is infinite variety in how different species respond to light, from the ranks of sweetcorn gobbling it up in their summer glory to the fern at ease in some damp rocky cleft or making itself at home in the low strip of shade along one edge of one of my polytunnels. That's not good practice, but not so bad as the examples occasionally seen of algae-encrusted tunnel covers under the murky light of which tomato plants languish as if in a dungeon – torture for plants! Something less easy to fix is the depressing flush of weed that envelopes your potato ridges in a trice once the tops have gone down with blight, and well before you can harvest the crop. You thought your spuds were pretty clean, but the weeds were there all along – feeble and half smothered, but waiting only to be released into the light. This ability to tolerate shade for long enough to outlast it is one factor that makes our chickweed and so on the weeds that they are.

The most extreme example I ever saw of plants finding a way of satisfying their need for light was in the fogou at the Iron Age village of Carn Euny in the far west of Cornwall. Fogous are stone-built subterranean passage-like galleries of obscure purpose found at ancient habitation sites in Cornwall, and in the north of Scotland too (where they are known as souterrains). The Carn Euny fogou has a circular corbelled chamber to one side of it, entered through a low doorway from the passage itself. A little dim light gets in that way and a little more via a small hole through to the soil surface at the apex. When Jan and I were last there we noticed that the granite blocks of the interior, or the interstices between them, appeared to be clothed with a sombre mossy growth. But then we caught sight of a sudden flash of brightness, green gilded with gold, which came and went (we realised) according to the line of sight. By cocking our heads at a certain angle and rotating our view around the chamber it was possible to see each section of wall light up and go dim again by turn. What magic was this? Closer inspection of the plant seemed to show that its surfaces act like mirrors, concentrating the little light available into its rosette of tiny leaves. I have been unable to find out anything about these plants, apart from the suggestion that they are bioluminescent (i.e. themselves light producing) which they certainly aren't. Perhaps a reader in that part of the world could enlighten us?

Schistostega pennata – luminous moss, also known as Dragon's or Goblin's Gold. "The luminous effect is due entirely to reflected light. The protonema of the moss develops cells with lenses that concentrate the feeble light upon the chlorophyll granules; these absorb such wave lengths as serve their purpose, reflecting back the remainder in the direction of its source"- R.Hansford Worth 'Worth's Dartmoor' Peninsula Press, 1994.

NATURE NOTE 24

Two onion kind

Summer 2015

The Rev. C.A. Johns in his Flowers of the Field (1851, and numerous editions for decades thereafter) says of ramsons (or wild garlic, *Allium ursinum*) that "the flowers are white and pretty, but the stench of the whole plant is intolerable".

Recently I was replacing a fence that I'd taken down so that the hedge beside it could be laid. The bank there, like so many of our hedge banks after a century or more of neglect, spills down into the pasture and – the coarser vegetation being long overcast by sprawling holly and teetering hazel – grows a fine stand of spring flowers. By mid-May there is a happy association of glossy white ramson and the sumptuous depth of bluebells, with here and there red campion and yellow archangel to set them off. My trampling as I unrolled, strained and stapled the barbed wire back to its stakes gave scope for the ramsons to release their odour of garlic. I thought little of it then but later, going to bed, found that it had impressed its pungent memory upon my mucous membranes. This was perhaps because, as it turned out, I was gestating a head cold – but this was a mild and earthy garlickness and not unpleasant. In any case being able to have the freedom and opportunity for immersion in the infinite variety of this inexplicable thing called Life was not the least of the reasons I became a farmer, so how could I complain?

It's likely though that if this was a hundred years ago (when Johns' estimable book was still in print) the scent of garlic would sit less easily in my nostrils. Tastes change. We're all used to garlic now, some of us even try to grow the stuff, but there was a time when unfamiliarity bred contempt. Even forty years ago when my wife Jan, newly here from Canada, went to buy some at a market stall in Exeter she was told "we don't sell that foreign rubbish here" (apparently foreign oranges, bananas and so forth were acceptable).

It's an interesting and attractive plant – "white and pretty" says little about it. The individual flowers, star-like and glistening, are arranged a dozen to twenty in loose umbels on top of stiff and curiously formed stalks. Generally described as triangular on close inspection the apex of the triangle can be seen as flattened, forming a narrow side opposite the widest one which is itself slightly dished. Its concavity results from each of the medium sides extending past it into narrow ribs, an irregularity that becomes tangible when the stems are rotated between thumb and finger. The leaves meanwhile are broad, smooth and shapely – archetypically perfect. They elegantly decorate the floor of ancient woods and shady places from mid-winter on, long before the flowers appear. Mild in flavour when cooked, they are similar in appearance to those of Lily of the Valley (which are poisonous). The bulbs meanwhile are loved by bears (and wild boars), hence the Latin name, but they have to dig for them. The plant produces contractile roots in the spring which pull them into the soil to a depth of five inches.

We are lucky to have another bulbous onion relative that beautifies our farm in May, one that is much less common and a good deal more exotic – the Star of Bethlehem, *Ornithogalum umbellatum*. It's probably native, but is also cultivated as a garden flower, the bulbs being available for purchase. Here it occurs mainly in grassland that was formerly orchard (going back a hundred years or more), an association suggesting that it may have been planted. Why anyone would plant it there is a mystery, especially as the foliage is reputedly poisonous. I don't let that worry me – the flowers appear before the cattle have eaten their way to the pastures in question and by then the foliage, such as it was, has died away to nothing. Before being overtaken by the grasses among which it grows it looks not unlike them, but the leaves are a deeper green than any pasture grass, more fleshy and somewhat glossy, with a silvery-white stripe along the length of each one. It's only when you've lost sight of the foliage, and only when the sun is shining and not otherwise, that the flowers are suddenly there, loosely clustered constellations of – you've got it – stars! The eye registers this with a kind of jerk. Unless you are actively looking for them there is always some sense of surprise, both because of the way their whiteness strikes the eye and because they float almost insubstantially among the workaday herbage around them, like visitors from another planet.

These indubitably star-like flowers are similar if not identical in structure to those of the ramson, and perhaps those of every other onion kind, but at an inch or more across are twice the size. The flower (technically speaking) is formed of three petals and three sepals placed alternately. The sepals are a little longer and broader than the petals, a difference that enhances the starry effect. The glistening of the ramson comes from its glossy and reflective surfaces. The Star of Bethlehem shines too, is almost dazzling, but while there is a sheen to it, its whiteness is deep and still, its light seeming to come from an inward glow. Each petal and sepal is backed with a broad greyish-green stripe so that when closed up you just catch a glimpse of white between bars of green. The source of all this excitement – the bulb – is said to be edible, but as they retail at £4-20 for ten you'd probably be better off eating shallots.

Some hardcore natural history

Autumn 2015

I dream of concrete but wake to hardcore. If you are not prepared to use herbicides and yet wish to impose some order on the space in front of your shed then concrete is your only man. In our early years here, fetching bags of cement in the van and aggregate from the quarry down the valley by tractor and trailer, I mixed a good deal of the stuff in our capacious but somewhat eccentric PTO driven Teagle mixer. But it was enough to get the inside of the shed surfaced and then a little projection beyond it on which to wash leeks and so forth. Ready-mix was not an option, even had it been affordable, because of the then impossibility of negotiating our access lane with anything bigger than a two-ton trailer. So for the rest it had to be hardcore which I could fetch from somewhere myself. Ironically this included the now redundant and broken-up first load of ready-mix ever to be laid in the village, but the bulk of it was unlawfully retrieved overburden from a disused roadstone quarry a short but tortuous distance away. I brought this home and bashed it up, where necessary, with a sledgehammer. As a free alternative to Terram ground separating membrane I used opened-out dumpy bags (then just beginning to replace hundredweight manure sacks) over the subgrade of rocks and subsoil excavated from the shed site. It's important to prevent the two layers mixing, as they otherwise surely will with time and traffic – the hard going down and the soft coming up. What you can't do is prevent the steady formation of a growing medium on the surface, a process which, as suggested in a previous Note, will eventually do for our network of trunk roads once the Highways Agency (or its successors) has given up the ghost (as it surely must).

Nature abhors a vacuum, a cliché whose truth is all too evident to any organic grower struggling to keep their crops clear of weeds. Concrete may take many human lifetimes before the vacuum is conquered, but hardcore presents no such challenge to natural forces. In former days of intense activity on this farm, with a constant succession of link box loads coming into the shed and van loads leaving it, the traffic was enough to keep the growth of vegetation to a minimum. Now activity has subsided to a shadow of what it was and growth proceeds apace. The stimulus for this is actually implicit in the previous traffic, given the ability of tractor tyres to pick up soil from where you'd rather it stayed (the field) and drop it off – complete with seeds – where you don't want it at all (the yard). About three years ago, fed up with the ragged-looking mess and faced by an event that required space for cars to park, I scraped over the area and spread 15 tons of scalpings on the surface. Smart! But the effect was short-lived and is now barely evident.

Growth begets growth. The due process of soil formation out of rock – wind, water, frost, lichens, mosses, a little accumulation of organic matter resulting and thus eventually space for flowering plants – is not a requirement for the overcoming of hardcore by plant growth as the material inevitably includes a proportion of dust and the rudiments of soil. Pretty soon there are struggling grasses and broad leaved weeds, each returning its hard-won organic matter to enrich the mix. Close to the hedge where implements stand and where there is the odd heap of sand or aggregate left-overs, but where any actual traffic is pretty much absent, the growth now is such that I sometimes feel compelled to mow or strim it back. Here in this suddenly wet late summer are docks and dandelions, little swathes of pretty medick (a kind of trefoil), some vibrant clumps of wild white clover, tangled spreads of creeping bent grass and the inevitable stroyle (couch). The broad pink flowers of a common mallow provide a shock of colour (I mow round this) while some spires of now faded purple linaria have outperformed the one specimen I allow space for in our herbaceous border. That plant was given us by a friend with a label describing it as "a weed sold in garden centres". Is it evidence for Rupert Sheldrake's theory of Morphic Resonance that others spouted very soon after from a recent addition of hardcore?

Closer to the shed front, on what is effectively a still-used track, are found the more typical plants of the frontier zone between natural exuberance and human-induced sterility, those that colonise gateways and farm tracks everywhere. Chief among these are broad-leaf plantain (our gift to America, and elsewhere) and pineapple weed (America's gift to us). You'll know this one I'm sure – a petal-less version of mayweed or chamomile with a strong and not especially pleasant smell of pineapple. It's not that these plants necessarily like compaction below and traffic above (I can show you much finer plantains in my tunnels) but they tolerate it, and – given the inability of plants to move around entirely at will – what they can tolerate must be instrumental in their success as species. As well as these two this poorer zone also provides a sort of living for some stunted grasses, chiefly annual meadow grass (that tedious weed), several dingy specimens of clover and some knotgrass (not a grass), which last can extend for two or three feet from one short and extraordinarily tough stem. A brighter note is chimed by a little pink-flowered cranesbill (probably *Geranium molle*) which has found a niche right where the hardcore butts up against the concrete.

The typical bird of weedy places is the goldfinch, a sign of a poor farm and/or of a poor farmer. Despite that I like to have them about the place as they are cheery, companionable birds. But they are only incidentally birds of gone-to-seed hardcore. The one that does typify this small habitat that I have, somewhat to my chagrin, brought into being is the dunnock (or hedge-sparrow). All through the winter months a small party of these gentle, mousey, unassuming birds spend much of its time fossicking about among the weeds and stones pecking up the seeds, and any insects they may find. For creatures that weigh only a few grams, and certainly less than an ounce, it provides a rich enough harvest.

B is for beaver?

Winter 2015

'Acorn, adder, ash, beech, bluebell, buttercup, catkin, conker, cowslip, cygnet, dandelion, fern, hazel, heather, heron, ivy, kingfisher, lark, mistletoe, nectar, newt, otter, pasture, willow'.

These are words that have been excluded from the latest edition of the Oxford Junior Dictionary to make room for the new additions – mainly those concerning digital media, such as "cut and paste". This is a first dictionary, aimed at seven to nine year olds, the number of entries is limited and the business of selection is obviously not straightforward. Certainly you can have some sympathy for the difficulties faced by the compilers. As innovation is an accepted good (the economy depends on it) there is bound to be a jostling for position, and where there are winners it seems there must be losers too. Perhaps now that smoking is a social ill, children no longer put the stalk of an acorn cup in their mouths and pretend to be puffing on a pipe and health and safety would rather they were not playing at conkers. But the acorn gives rise to the oak, the one tree that the English ought to be able to recognise and even when just held in the hand the smooth and lustrous conker is a thing of pleasure. However you look at it, it is a sorry tale. Each of these words is bright, instinct with life. Each speaks of beauty and interest of form, of the simple grace of unforced existence. Taken together they read something like a hymn, emblems of an earthly paradise.

From acorn to willow they memorialise the speech and perceptions of our forebears, the naming of things being at the heart of humanity's relationship with the world in which it lives. A name is not just a useful designation, it is also an expression of the essence of that name's possessor. What is nameless can be no more than insubstantial, shadowy and even sinister. I will leave you to reflect, if you wish, on what these omissions say about where we are in our relationship to the natural world and the future of the human species.

Outside the pages of the Oxford Junior Dictionary the otter has been doing pretty well lately and is a good deal more common now than when I was a young man. This animal has a particular resonance in Devon, largely because of Henry Williamson's Tarka but also because we have a (by our standards) a substantial river named after it. The valley of the Otter has some of the best farmland in the county – rich pastures (if I may use that adult word) and good light arable soils too. With moderate rainfall and notably warmer than average summers it is one district in the Westcountry where you can be fairly sure of growing a decent crop of sweetcorn. It seems an odd choice then as a place to establish a colony of beavers, but this is what has been done and I understand that the project's instigators consider it a success.

The beaver was once indigenous to this country, to be found for instance at Beverly in east Yorkshire. Though lingering along the Teifi in Wales until into the thirteenth century, in England it seems to have died out in Anglo-Saxon times, a victim of the same human population pressure and consequent demand for agricultural productivity which did for wolves and led to mediaeval deer being enclosed within park pales. Delightful creatures that they no doubt are, beavers are capable of extraordinary feats of landscape engineering which may not be always welcome. The one time I walked across a beaver dam I found it almost impossible to believe that it was not the creation of a gang of civil engineers equipped with chainsaws and earthmoving equipment. This was a longish time ago in Ontario and the details are inevitably hazy but I remember a shallow, loosely-wooded valley. The river, a little creek by Canadian standards, had been choked off by a dam maybe thirty yards long, six feet high and twice as broad, the top being wide enough to drive a jeep over. Nor would there have been any trouble in driving the jeep across it, such was the solidity of its construction. A prodigious number of trees must have been felled and somehow floated into position, hence I suppose the relative openness of the surrounding bush, but also (it seemed) a very considerable quantity of mud gathered up with which to caulk it. At any rate the structure was completely stable and well grown-over with vegetation. The area of creek bank and hinterland affected, taking into account tree felling and disruption of drainage, while of little significance in Ontario would be capable of making a significant dent in Devonian agricultural production, assuming that "success" means that the Otter beavers will increase, multiply and extend their territories into other river systems.

Such introductions made and proposed under the banner of re-wilding, with the implicit sense of doing right by the natural world, seem to me to be in part a kind of self-indulgence. If beavers could not outlast the pressures of life in Saxon times, can they can be accommodated in a countryside a good bit smaller than it then was and with a human population now forty times greater? Instead of pursuing what is likely to prove a freak-show rather than a genuine re-naturalisation, we should be cherishing what we do have, the mostly common and unremarked animals and vegetation – just such as those listed above – that have so far survived and even prospered alongside long centuries of human activity. There is threat enough to them in our ever more rapacious system of agriculture as well as in the indifference and lack of comprehension on the part of the consuming public. Ditching them from a children's dictionary, perhaps a small thing in itself, can hardly be a step in the right direction.

I'm now quite looking forward to there being beavers in this valley, as is likely to become the case in time. Friends that moved from here to the valley of the Otter think that being able to see beavers is among the best of its attractions. Their re-naturalisation is even said to have boosted the local economy. I could have been more open-minded, but the general point stands.

NATURE NOTE 27

Animal spirits

Spring 2016

Organic growers and farmers often characterise what they do as 'working with nature' and leaving aside philosophical questions as to what the word 'nature' encompasses and what's natural and what isn't, that's a fair statement of our approach to husbandry. The relationship can feel like an abusive one though. Perhaps that's one dividing line between organic and non-organic agriculture – on which side the abuse is dealt out.

In the days when I used to drive a van-load of produce at least weekly to the co-op warehouse I more than once found myself behind a car with a sticker in the rear window proclaiming that its occupants loved wildlife. "Huh, not me" I'd think to myself, mindful of the destruction to our crops wrought by badgers, pigeons, voles, crows, squirrels and rabbits, among others. These days I'd have to put deer at the head of the list, but they are a relatively recent arrival.

We are lucky with rabbits on the whole, usually only suffering a bit of damage around the edges. Those few we have can be pretty irritating however, as (for instance) when not liking the taste of the annual flowers that we planted in among the field crops they would just dig them up and cast them aside, seemingly out of devilment. But describing this as "devilment", besides being over-indulgently anthropomorphic, begs a question. While, as has been recently announced, we can detect minute gravitational waves at the limit of the so-far envisaged universe our knowledge of the mental workings and motivations of the creatures we share the earth with is precisely nil. Forget "Peter Rabbit", the mind of a rabbit is a closed book! And what about the cock chaffinch that spends a fair part of each day pecking at the wing-mirrors on our parked car, speckling the paintwork with his droppings while he's at it? He's very determined, only reluctantly flying to a nearby low branch as I go by and soon returning to the task, or entertainment. Maybe he really is deluded into thinking that the mirror holds a rival to his space, but it is possible that some other take on reality than our own is at work.

As motiveless crimes committed by wildlife go (there's that anthropomorphism again) a recent inexplicable event takes the biscuit. We have a wheeled digger which through the winter is used mostly for feeding round bales of silage, scraping out the bullocks' feeding area and moving straw. For these purposes I start it up every second or third day and open the engine cover to check the oil and water every second or third week. To avoid having to unrope and uncover the sheet of the main stack too frequently I transfer a stock of big straw bales to the back of the fourth bay of the cattle shed at intervals and the digger gets parked in front of them, the only place it fits under-cover. The cattle occupy the other three bays. Most winters there are signs of a rat or two in the main stack, but as we don't feed our animals anything other than grass silage and don't currently keep poultry there's only a bit of stray grain in the straw to encourage them. This year is the first time their presence has been evident in the shed itself.

The last time I opened up the engine of the digger I was astonished, nay stupefied, to find that it had been entirely replaced by a rectangular bale of straw. How could this be? Had the world turned upside down? Recovering from this momentary stupor I saw that the engine was of course still there, the stub of exhaust outlet and air intake being just visible, but the entire space between engine and cover, from the back of the radiator to the front of the cab, had been packed firmly with straw. It took me several hours to clear it, borrowing a compressor so that I could blow out every last bit from in among all those nooks and crannies with which an engine and hydraulic system abound.

A few things became obvious. The straw had been brought from the edge of the bullocks' lying area – where not dried out by proximity to the engine it was damp and mucky; there was no trace of rats or of any purpose to the activity – no runs, no nests, no shit; it was lucky, to put it mildly, that the digger hadn't gone up in flames, possibly taking the shed with it. Much more remains inexplicable. There is no evident way that the straw could have been elevated the 18 inches from the ground to the engine compartment. A great deal of dedication seems to have been expended on working the material in and around wires, exhaust manifold, fuel pipes and injectors and the mass of hydraulic hosework just below and in front of the windscreen. Was the work carried out all of a piece, or done bit by bit between the machine being used? And what could have done it?

I'm assuming it's rats, because I know they are present and cannot see what else could be responsible. A neighbour suggested that I might have annoyed the little people, but I'm careful to do no such thing and in any case pixies seem to have died out in Devon in about 1950 (leastways – I've never seen any). If it was them we could put out bread and milk to keep them sweet (not good for hedgehogs, but fine for pixies). But of course it's rats – what else could it be?

As to why . . . who knows? But I'm now parking the machine well away from any straw.

NATURE NOTE 28

Summer skies

Summer 2016

Late May, the day soft and warm, approaching sultriness. Cloud drifts aimlessly, unthreatening, just patching the sky. The rain was yesterday – today the sun's hazy heat on moist ground and sap-filled vegetation brings recognition of what summer is and remembrance of all those other summers lived. Is it strange that the seasons cannot really be known until each one comes, touching the skin, reaching into the body's veins and sinews? In winter the coming summer and all summers before it can be appreciated, their nature understood, but only by the mind as an abstraction. The reality is foot-slogging mud, or frost freezing the fingers as you make what you can of the few hours in which summer-grown crops can be prised from the earth. In winter's thin air summer has no more meaning than an aspiration! Then summer comes. The air, now weighted with the scent and exudations of leaves beyond numbering, becomes somehow tangible and presses on the senses, and winter in its turn is past. The body forgets it, its inner reality becomes unknowable.

Towards dusk the undersides of the clouds turn milky grey as the sun dips below the hill, a different hill to the one that shuts off the brief day of winter. The angle of light striking the clouds' edges beads them with gold or silver, emphasising the purest blue of the spaces between. The day lingers into stillness in which sight and sound take on a special clarity. From somewhere near the top of a still unfurling ash a thrush colours the air with a song that somehow crystallises the heart-stopping beauty of the newly minted world. Its song is so sweet, so perfect, so delightful to the senses, that you'd like to fold this moment around like a parcel and carry it with you forever. But of course you can't do that. What you can do is stop and open your being to the moment. These days you can take neither song thrushes nor ash trees for granted, and it demands nothing less.

The sun now sets north west, so that the last of the light leaves us to the north. Opposite and rising with the night sky in the south is a notable grouping of stars. Some way above the horizon and in the constellation of Libra, unmistakably and thrillingly red, is Mars, one of the two brightest objects of the June sky (the other is Jupiter, going westward and setting an hour or two after midnight). A few degrees to the left of it and on the same level lies Saturn, not so bright but still conspicuous. In this position Saturn is a spectator on an age-old quarrel of the heavens, for below it is the constellation Scorpius, which straddles the southern horizon for the few summer weeks during which it is just visible. Astrologists say that Mars is the ruler of Scorpius. Yet the chief star of Scorpius, its glowering orange-red eye, is Antares – itself one of the brightest of summer stars. Although not far above the horizon it should be easy enough to spot given a clear line of sight and not too much light pollution. Antares is 'anti-Ares', Ares being the Greek name for Mars. It is the rival of Mars, its eternal competitor and there the two are, doomed to be in close and uneasy sight of each other all through this summer. What I know about astrology isn't worth knowing, but it seems to me that this conjunction of enemies, both pulsing with aggressive energy, must betoken a deal of turbulence for us self-poisoning Scorpios.

No such baggage attaches to another glory of the spring and summer night. If you follow the line of the last two stars of the handle of The Plough a short distance across the higher reaches of the sky's dome to the topmost stars of the next reasonably obvious constellation, which is Bootes, and then drop your sight a little to the south and west you won't fail to spot Arcturus, the chief star of that constellation, third brightest in the sky (after Sirius and Vega) and one that should be recognised by all of us who cultivate the land. To agricultural societies Bootes is the ploughman, his hand to the plough, though in other contexts he has been seen as the hunter or keeper of the two bears – Ursus major and minor. In our latitude Arcturus rises to brilliant prominence in early spring, just as the drying winds that come with March allow us to get onto the land and set the whole cycle in motion once more.

Our summer visitors are back safe and sound. The flycatchers returned to find that the tall and gangling half-wild plum they nested in last summer is no more. They might have thought of returning to the implement shed where they made their home for several years, but this was blown apart in an exceptional storm from the north in early March. I see them perched in prominent locations round about, from where they dart out for flying insects in that characteristic way of theirs – so they must be nesting somewhere. I've now netted off most of the loft of our pack-house shed (why didn't I do it long ago?) so that the swallows can continue to nest in their part without fouling the rest of it. They seem happy enough with this arrangement, but it remains to be seen how the fledglings will cope with netting hung close by when first they leave the nest. Downstairs, between loft floor and the boarded ceiling of what used to be our office, there is a space that is stuffed with wild bees. They go in and out via a hole between the top of the window frame and the cladding above. All day a mass of bees jostle and buzz around this entrance. At night when they are gathered in the hum is of a steady dynamo. Here we have the advantages of bees without the business of keeping them – except for one thing. Maybe I could drill a hole to let the honey out?

Mind that tree

Autumn 2016

In the early spring of last year I planted three ash trees. As it's currently illegal to move young ash trees from place to place I should explain that they were self-sown seedlings which I dug up, carried a few yards and transplanted where I wanted them, just up from the ford which is this farm's back entrance. Now in the third year of their lives they are chest high, clear of their plastic guards and beginning to wave about in the air a bit. Planting any tree is something of an act of faith I suppose, but planting ash trees now that chalara, ash dieback, is abroad in the land is an act of faith indeed, or an act of folly. The last I'd heard there was one outbreak in the county and that at some distance from us. Now looking at the Forestry Commission's map that traces the disease's advance I find that our ten kilometre square has been coloured in this summer, along with most of the neighbouring ones.

Directly across the track from the ash I transplanted an oakling which had sprouted inappropriately in a nearby gateway. There's a symbolism here. At what is very nearly the top of the farm there stood side by side, when we came here, a fine ash and a fine oak, evenly matched, their branches mingling in a happy marriage. The entrance lane, formed with pack ponies in mind, made a right angle turn to pass between them before taking its sunken course down to the old house and yard. When looking up from the yard they stood as sentinels etched against the sky, and were visible as such from much of the farm; coming down to it you passed through their imposing gateway. But the ash lost most of its top in a storm one now distant winter and then the oak was felled by an evil blast of wind about five years ago. You try not to read too much into these things – disasters will happen, and there was some convertible timber and a great deal of firewood to compensate – but we were left stunned and bereft by this second blow. However there's a youthful ash nearby that will almost stand in for the old one. As for the oak – I prised out a seedling that had sprouted beneath in the stony bed of the old lane, and planted it nearby in the hope that one day the marriage will be restored, though quite how the old folk of the farm contrived to get evenly sized ash and oak trees to stand together in this way given their very different growth rates is a mystery to me. Or was it just a chance of nature?

So at the ford – some years hence, with a good deal of luck, and dieback and other perils permitting, there'll be an oak on one side and one of the three ash trees (whichever is fittest) to accompany it, arching over the track in another happy conjugation, at one entrance as at the other.

Trees of an age best suited for transplanting are cheap enough – the guards and stakes and weed-suppressing mats cost more. Planting is quick, especially for a grower used to this kind of activity, but for the first few years some care, to keep the competition at bay, is essential. There is thus a small outlay and some work, but most of all a need for forethought. I'm sure all growers on occasion plant things too closely together so that the whole suffers from lack of light and space, and access for weeding and harvest becomes irritatingly difficult. The smallness and apparent fragility of a young tree in the hand makes it hard to visualise the dimensions of the giant that we hope it will one day become. On a livestock farm some loss of grazing yield may be a fair trade for the shade and shelter that a well-grown standard tree provides but where vegetables are grown the competition for light and moisture can be significant.

Ash trees are notoriously demanding of soil water – here in the grip of a prolonged drought the spread of their roots can be plainly traced in the arc of whitened grass extending out from the hedge. I remember the effect on a field crop of french beans in a previous droughty summer, where the course of individual ash roots reaching out into the body of the field was marked by wilted bean plants. When it comes to shade I discovered that the effect is more than a simple interruption of light. There was a tall overgrown hedge on the north side of one of our polytunnels but some yards from it. Despite the lack of direct shading the effect of additional available light in the tunnel was marked once this hedge had been cut and laid, as if the trees had somehow previously gobbled up much of the light before our plants could make use of it.

Where new trees are wanted there may well be seedlings present which just need a bit of light and space, and freedom from the dreaded strimmer, to have the chance to make a better tree than will result from any introduced transplant. Often, probably usually, planted trees come from stock that is not native to the district or even the country. Despite being ostensibly native species they may show characteristics which are decidedly foreign. An example of this is a linear planting of hawthorn beside a new link road on the edge of town. Admittedly this is a warm spot where the blackthorn manages to bloom long before it does in this valley, but even so there is something suspicious about whitethorns that are always in flower before Mayday, as these are. I suspect the landscape contractor put in stock of distant origin, it being cheaper to import the plants from, say, Hungary than to use home-grown material. Such activity, driven by convenience and cheapness and – due to WTO agreements that prohibit restraint of trade – impossible to disallow until too late and the harm is done, underlie the threats to the integrity of our countryside. Chalara was first identified in Eastern Poland in 1992. It took just twenty years to get here. Whatever next?

All three of those ash that I planted succumbed to the disease by their fourth year. Misguided faith, or pure folly?

NATURE NOTE 30

Hip, hip . . .

Winter 2016

There's a field gateway here, we call it the Birthday Gate, where there has been a fine display of autumn berries. A large, ivy-clad ash stands a few yards along the bank from the latching post and beneath the reach of its branches grow four out of our five common hedgerow fruits (you have to go on a bit further to find a sloe). Closest to the post and out of what is really more of a heap of rocks than a hedgebank, a wild rose clambers up the remains of a dead spindle bush, climbing high above to show off its lusciously plump and scarlet hips in plain sight. Next along the hedge is a small hawthorn. Over-topped by the rose it never produced much of a crop of haws, and the few that the birds have left are now darkening with age. The remaining space between it and the ash is filled by the curious coupling of a spindle and a holly.

These two have rooted a few inches apart no more than a foot away from the base of the ash and grown up together in a mutually supportive tangle of limbs. By this means the spindle, usually not much more than a bush, has got itself twenty foot and more up into the lower branches of the ash where its fruit shine out against the darkness of the holly and ivy leaves. The holly meanwhile fills a half circle around the ash's trunk and sprawls outward on the lower side, overhanging the fence that reinforces the bank at that point. It had a grand array of berries, but the birds took all these as soon as they found them ripe enough to eat, quite early in November. There is a good deal of diversity in hollies – some for instance making stiff upright growth, others sending out long sprawling shoots that root again where they touch the ground. By the same token some trees get stripped as early as October while on others the berries are left alone until they become palatable in the new year. Evidently there is more than just redness to their being ripe enough to eat.

It's easier to recognise ripeness in a spindle. The fruits can seem almost impossibly exotic for an English hedge and are all the more startling for being borne by a bush that is otherwise easy to overlook. It makes itself obvious only when the berries, which are four-lobed purses, each lobe bivalve and velvety like a tiny peach, turn at the onset of autumn from green to a unique shade somewhere between pink and cerise. Then it becomes unmissable even at a distance, the more so in that it is not shy in bearing so that every one of its thin twigs is bowed down with the weight of fruit. In November the lobes split apart along their bivalve creases and gradually open out so that the bright orange fruit within is revealed. Resembling a small lemon pip in size and shape, technically this is an aril, a fleshy envelope for the hard, white seed within. Once fully visible the birds start to take them, though they don't appear to do so with quite the same gusto that they apply to haws and holly berries.

There is a hollow in the base of the ash from one side to the other, though in other respects the tree seems sound enough. It would be possible to crawl through it if the narrower end wasn't blocked by those twin stems of the holly and spindle bushes. As it is it makes a useful shelter for probably a variety of creatures. At one time it was a fox's lair, judging by the bones and mess left within. Just now the interior is clean enough, layered with a soft dry dust of decayed wood and vegetation which is scattered with the pink husks of spindle berries, glowing gently in the gloom. Obviously it's not just birds that eat them, here some small rodent has had its fill. The spindle also has an insect connection – one that might be thought to be of particular concern to growers. The black bean aphid, that especial terror of broad beans, overwinters as eggs on spindle bushes specifically. Sometimes it is suggested that they should be rooted out on this account, but we have a great number of them scattered through the hedges and wild corners of the farm and we don't seem to have suffered unduly from the black army because of that.

We can't eat spindle or holly berries – for humans they are both poisonous to a degree. Haws are edible, their scanty flesh mealy and a little sweet, but they seem barely worth the effort. Rosehips are dry too and unpalatable due to the seeds within being armed with stiff pointed hairs which are highly irritant if digested, and so far as I can see birds do not take them. But they have four times as much vitamin C as blackcurrants and twenty times as much as oranges so the effort of turning them into syrup, the seeds held back in the jelly bag, is well worthwhile – and it's delicious too, very good with yoghurt. Between 1943 and 1945 voluntary pickers harvested an annual average of 450 tons, some of the resulting syrup being earmarked for mothers and children at subsidised prices. Thus was the health of the nation during wartime enhanced by this freely available hedgerow fruit. Less edifyingly – in my boyhood a crumbled rosehip shoved down the back of someone's shirt was considered as effective as any purchased itching powder. And so it would be as at least some propriety itching powders (how many can there be?) still consist of exactly that.

NATURE NOTE 31

A sense of purpose

Spring 2017

It's warmed up lately (mid-February) and so it was no surprise, but
still a welcome sight, to see that the frogs have deposited their bit
of spawn in its usual place. This is a stretch of still water about ten
yards long and two or three wide at the lowest corner of the farm,
lying close and parallel to the brook, and at a slightly lower level. It
occupies part of the brook's former channel, the course changed
perhaps by a fallen tree damming the flow long before our time here.
That's all it takes, succeeding spates pushing up flotsam and then
dislodged stones against the straddling trunk until at last the pent-up
water forces its way over the upstream bank and cuts a new channel
as it goes. At the top end of this backwater is an elongated depression
of soft ground and marshy vegetation, the remains of the top end of
the old channel, out of which a trickle of water flows for most of the
year. At the lower end the overflow seeps out into the brook's new
course, but there is no direct connection between the two so that this
bit of water makes a usually safe nursery for the hatched tadpoles.
It's not an attractive one to the human eye. The water is perhaps
eighteen inches or more deep, but the bottom is thickly obscured with
a broth of rotting twigs and last year's leaves, darkly mottled in its
slow decay. Beneath the surface – where the pond skaters are already
gliding about and which later will be busy with whirligig beetles,
whizzing around like aquatic dodgems – the water looks quite lifeless
to the naked eye. But it must be full of sustenance – at any rate the
tadpoles seem to do well enough, though it's hard to be absolutely

certain on this point once they've spread out from the spawning site in the shallows. Shade from the trees round-about make it easier to see the web of sallow and alder branches above that are reflected on its surface than any of the life in its depths.

One spring the spawn was washed away in a spate, but other than that this site has never yet failed. There are a couple of other places on the farm where spawn can be found but these are only winter-wet and need plentiful rain in spring and early summer if the tadpoles are to have a chance of survival. You never know with frogs. Mindfulness does not seem to be part of their makeup and they sometimes deposit spawn without any apparent concern for the future of their offspring. I once found a considerable blob at eye level on top of the horizontal limb of a nearby tree. The common frog is said not to be a climber and I have no idea how it got there. It's one of those small wonders which the natural world provides. Out on the moor in March extensive deposits can sometimes be seen lying where there is little or no water even then, and where there certainly won't be any later, even though there are plentiful hollows close by that will remain reliably water-filled but which the frogs have ignored.

It's curious to look across the dark reflective surface to the brook just beyond, its surface bustling through crests and hollows as it hurries over the stony bed, making white water as it does so. It's an odd juxtaposition – you half expect the vigorous stream to leap out of its shallow channel and overwhelm the quiet pool beneath it. Sometimes winter flooding does give the pool a bit of a sluice-out but up to now the separation has always been restored and so it continues to provide the frogs, and no doubt much besides, with a useful nursery. Who knows for how long? Geomorphology – the shaping of landforms – is at work here, if in a small way. Along the upstream fields where the brook marks the farm's boundary it is largely confined by stone walls (massive, moss covered and impressive even in decay) or, set back a bit, by hedge banks of earth and stone. Here and downstream where it flows for a stretch through an untended wilderness there are no such restraints and, still retaining pent up energy from having fallen seven hundred feet in two miles, it has achieved a good bit of shape-shifting in recent wet winters (but not this dry one). Its filled in pools, decommissioned waterfalls (and made some new ones), spread fans of stones and then gravel, entirely islanded a substantial clump of alders that was formerly on dry land, and so on. Always tending to spill and multiply at this point, new channels have been made, old ones reopened and more recent ones blocked off. Never mind geomorphology, this activity has administrative and legal implications. Admittedly these are hardly earth-shaking – a matter just of the brook marking the line of the boundary between two parishes. Currently I'd say that ours was on the losing side.

The sallow is quick to colonise wet places and has great agency in changing landscapes. Alders like wet too, they need it more than sallows do, but they tend to grow fast and straight until they simply topple or fall apart with little in the way of branch-wood to cause an obstruction, and then soon rot away. The sallow on the other hand produces a confused mass of spray from a confusion of branches. Still fully rooted the trunk leans down and lets its limbs split and bend without fracture until they can touch the ground, whereupon it quickly strikes new secondary roots. It's primed also to vigorously sprout upright shoots from limbs and branches that formerly were vertical but now approach the horizontal. In these ways it is a great obstructor of waterways and of the paths beside them. Both tree and frog exhibit a wonderful fecundity, but in this the tree appears to show the greater single-mindedness of purpose. The brook meanwhile takes what opportunities it can with no evident guiding principle other than gravity.

Sallow is a general term for the broad leaved Salix cinerea and Salix caprea – the grey and goat willows. Both are pussy willows in springtime. Alders are locally known as Allers.

Invasive plants

Summer 2017

From the Birthday Gate, mentioned a couple of notes back, the pasture field slopes down to a roughly level space and then dips away again into a sort of linear sump bounded on its further edge by the bank, part earth, part drystone, which borders the brook. The drainage water that emerges here from little springs and seepages can only find its way into the brook someway downstream and so this area is permanently damp and in a wet year, unlike the one we've just had, becomes impenetrably soggy. Sallows sprawl across it and alders totter from the bank. It's a place of aquatic grasses and sedges, with golden saxifrage in springtime and wild garlic and yellow archangel on the slightly drier spots, and – where less shaded – lady's smock, ragged robin and spearwort, the buttercup of marshy ground. Looking down from the gateway in late March my eye was caught there across the intervening hundred yards by a garish splash of yellow – the emerging spathe of a skunk cabbage, *Lysichiton americanus*. It first appeared last year but investigation now revealed two more plants also soon to flower and seed. Obviously related to our native cuckoo pint, *Arum maculatum*, but true to its specific name an altogether larger and more showy plant, it's cultivated in bog gardens both for that striking yellow spathe and for the three feet tall glossy and crinkled leaves that follow it. First recorded in the wild in 1947 it has since exhibited an alarming ability to take over the flora of damp woodland. Come Easter Monday, and with this in mind, we descended on the three visible plants with shovel in hand and dug them out. As for its suggested skunky smell, we caught no whiff of it. Perhaps the odour of upturned deoxygenated mud and trampled water mint was sufficient to overlay it, but if so it doesn't say much for its skunkiness.

It's not hard to see how these plants arrived. Some upstream neighbours have it in their waterside garden and a not so distant winter flood evidently carried its seeds downstream to lodge and germinate here. At one time I used to see a lot of seedling Himalayan Balsam, in similar habitats close to the brook and was assiduous in pulling it out. No plant that I know of has a more tenuous connection with the ground or is easier to uproot. Even so you might think I'd have been better employed cultivating our vegetable crops, but I've never much liked Busy Lizzies and while this taller relative's exotic form and colour may be welcome in areas of industrial dereliction I didn't want it taking over the native flora of our pastoral valley. This is the problem of introduced plants – it's not what they bring to a scene but what they subtract from it. It seems the source for this one has diminished or disappeared, at any rate we seldom see it now.

Another relatively recent arrival, at first close to the brook and latterly in fairly moist and shady spots at some distance from it (even including a rut well under the eaves of our cattle shed) is pink purslane, *Claytonia sibirica*. I first saw it in the early 1980s in the Tamar Valley, but both it and its salad-bag cousin, *C.perfoliata*, were recorded in the wild before 1850. Another North American native, it turned up on this farm before the end of the 1980s – a few years after we did. I can't speak for its edibility or otherwise, presumably you'd want to pick the leaves before flowering, but it's a handsome plant in a small way, with broadly ovoid leaves springing directly from the stem and five-petalled flowers of a more or less pale pink marked with deeper veins. It would be impossible to extirpate, so thickly does it grow in masses intertwined among the herbage native to its habitat. It seems to have found a niche to fill where it is not obviously over-competitive to the original vegetation, all the same it must now occupy sites that formerly grew celandine, yellow pimpernel and the like.

An altogether more aggressive species, but like the purslane originally ill-advisedly introduced as a garden plant, is alkanet, *Pentaglottis sempervirens*. This is perhaps the only plant with an English name derived from Arabic – "little henna" from al-hennah, the henna plant, through the Spanish alcanna – on account of the red dye obtainable from its roots. These roots are deep, fat and as tenacious as docks. It grew here close to the old house but lately has spread throughout the farm. The flowers are cheerful and attractive – five neat little royal blue petals springing from a bright white eye – and much visited by bumble bees, but in other respects the plant is a dull dog, with bristly nondescript leaves on two to three foot long stems. It's a bully too, overgrowing the familiar spring flora and continuing to occupy the ground for almost the whole year (hence sempervirens, ever-living or evergreen) so that once established nothing else has much chance. There's a quite frequent Maytime association here, a patriotic red, white and blue of campion, stitchwort and bluebells. To my regret in places the alkanet is hustling out the campion and stitchwort and replacing the bluebells, so that for all the pretty but transient blueness of its flowers the result is a dully homogenous uniformity.

Such changes are not solely the result of introduced species proliferating. I've heard dog's mercury, a native, described as the world's most boring plant – and even if rare it would be hard to work up much enthusiasm for it. For some years now it has been vegetatively expanding its dreary clonal drifts – usually all male or all female, the distinction is not exciting – both on the field banks and in our small patches of hazel coppice, steadily supressing the primroses, wild garlic, violets and bluebells that one would rather see. Why individual plants grow exactly where they do and why they wax and wane are matters that are hard to elucidate – variations within soil nutrients, shade, drainage and trampling by livestock as well as subtle changes between them are some of the factors that must come into it. To that partial list we now have to add changes in climate.

An elegant sedge

Autumn 2017

The outer edge of polytunnels, where the plastic meets the ground, is not an easy zone to manage. Too close with the mower or the hook and the plastic gets cut, left alone and a variety of plant life insinuates itself between skin and soil, some of which then grows up to deprive low-level crops inside of light. In the case of creeping thistle the rambling roots may find their way under the buried plastic, and once let loose inside it is the devil's own thing to control. It is a favoured place for nettles whose sinewy roots extend themselves for a distance in the tight space along the edge, throwing up groups of stems as they do so. Another frequent irritation is seedling ash trees, the keys lodged against the edge of the plastic by wind or by rain washing them down the side of the cover. They soon make sufficient dormant buds to regrow when cut so that only pulling them out will do for them.

It is a wet zone at the best of times due to the flow of rain water off the skin combining with the retarded drainage of the buried plastic. One edge in particular, due to the shading effect of the adjacent tunnel, is marked by damp-loving willow herb and even a couple of clumps of rushes. It was here a few years back that I noticed a plant, neither a rush nor exactly a grass, that suddenly caught my attention – though it must have been there for some time without my ever registering it. Out of a broad clump of narrow, shiny and rather lax leaves, deeper in tone than most grasses, arose slim upright stems bearing at wide intervals pretty if rather insignificant golden-green flowers. From the base of each flower grew out a long delicate subsidiary leaf, or leaflet, some nearly a foot long and all wavering around at a broad angle to the stem. This was unlike any grass I knew of and had not the roundness nor coarse pithiness of a rush. It had a special elegance of structure and composition, the fine arching leaflets contrasting nicely with the stiff erect stems and their understated studding of intricate little flowers. What was it?

Years ago as an agricultural college student I was required to collect and display in an album twenty (or more) different grasses, with notes as to their agricultural significance – if any. I have it still. You couldn't do anything at Bicton College without being marked for it and for this effort I was awarded 70% and the comment "Excellent collection". I'd have done better still if I hadn't mistaken a Poa trivialis for a Poa annua and a wood small reed for a reed canary grass. I like plants, but don't have the application to be much of a botanist. One I got no mark for was what I thought was heath grass. Against this the lecturer had written "This is a sedge".

Perhaps this attractive but previously overlooked specimen of the tunnel edge was also a sedge. There was nothing like it in my Observer's Book of Grasses, Sedge and Rushes, but then I found it in Keble-Martin's Concise British Flora in Colour – *Carex remota*, distant-flowered sedge. I felt quite pleased with myself – this was the first individual sedge that I'd ever identified! The long subsidiary leaves, it turned out, were not leaves at all (grasses, sedges and rushes don't have leaves like that) but bracts, though the distinction seems to me to be rather a fine one. The 'remota' refers to the unusual distance between the individual flower spikes.

I hesitate to tread in the area of taxonomy, a science about which I know nothing worth knowing and one that, into the bargain, seems to be liable to constant change. However I will hazard the statement that all the subjects of that Observer's book are members of the order *Glumaceae* with wind-pollinated flowers formed of spikelets enclosed in chaffy scales known as glumes. Of the three families of that order likely to be met with in our temperate climate the *Gramineae* (grasses) provide the greatest number of the staple crops – wheat, maize, millet, rice etc. – upon which so much of human life depends. *Cyperaceae* (sedges) and *Juncaceae* (rushes) on the other hand, while having some use for matting and the like, are largely inedible to man and beast. The only exception to this appears to be *Cyperus longus*, Sweet Galingale, an uncommon inhabitant of wet meadows in the south of England whose aromatic rhizome "was formally much esteemed as a tonic". Grasses tend to be juicy and sweet, whereas sedges are dry and tough. The simplest distinction, requiring no botanical expertise, is that while the former have hollow stems with solid connecting nodes at intervals, sedges have stems that are solid throughout and which tend to be angular, so that you can feel the difference just by running them between your fingers.

Since that discovery I have found *Carex remota* growing in several places on this farm where there is both shade and running water close by. Each year now I bring in a few stems to put in a vase and decorate a window sill. How could I have missed it previously? Wake up – there is little enough time left to take in the wonders of the world!

Astonishingly it is possible to buy plants of this sedge over the internet at £7-50 each. Meanwhile the German specialist perennial seed company Jelitto offers *Carex remota* (among forty other *Carex* species) at €144 per 100 gm (or €2 for a packet). It's an easy germinator and a hundred grams should give you fifty thousand plants. Are we in the wrong branch of the trade?

NATURE NOTE 34

Cabbage Whites

Winter 2017

Mid-November and the Large White caterpillars that infest the visible, aerial parts of the brassica crop have finished their feeding and just about, but not quite, disappeared from sight. Their last act as caterpillars is to undergo their final moult, revealing the hard case of the chrysalis in which they overwinter as pupae before emerging into flight as the first generation of next summer's butterflies. They need to attach themselves to some secure anchorage for their winter dormancy and to this end they are capable of traversing considerable distances through tangled grasses, weeds, green manures or whatever else lies in the way. From my own now small plot of brassicas it is a minimum of twenty paces to the nearest horizontal structure, a polytunnel. Here some have simply stuck themselves to the plastic skin, but as many again have found the end doors and have attached themselves to the timber framing (some outside and some in), often on the underside of a member where they are relatively snug as well as hard to spot – an attribute aided by their greyish casing giving little contrast to the background of weathered wood.

At the lower end of the plot, away from the tunnels, many have perhaps made for the nearby hedge, their chrysalides invisible among the forest of woody stems. Others have gone in the opposite direction and found the stakes that support the electric fence separating the vegetable corner from the cattle grazing in the rest of the field. They can hardly have wandered off through and over the grass simply in the vague hope of finding a suitable lodging place. Somehow they must have had sense that out there, several or many yards away from their feeding site, were solid rot-proof plastic surfaces that offered angled mouldings against which to lodge for greater security.

How many of these chrysalides will actually produce large white butterflies next summer is one question, but it's worth giving a thought first to what proportion of the caterpillars originally hatched make it this far. As all growers will know the Large White butterfly (*Pieris brassicae*) lays generous numbers of eggs in clusters of a hundred or more. The resulting caterpillars are strongly gregarious, especially in their early stages, and highly visible on account of their colouration and the exposed position which they occupy on the plant. This is quite unlike the Small White – eggs laid in batches of a few at most, the caterpillar active in the depths of the plant and camouflaged by its soft velvety green colour and by its habit of aligning itself along the leaf stems. The difference is down to the smell, or lack of it. Large White caterpillars concentrate the mustard oil that's in the leaves they eat so that they smell, some might say stink – and certainly the odour of infested brassicas is a distinctive one. But each caterpillar doesn't smell that much and its only by being in a crowd that the combined body odour is sufficient to send a suitably off-putting message to would-be predators. Another thing they can do en masse, which would be pointlessly ineffective if performed by a solitary individual, is a sort of Mexican wave. Maybe you've seen it? As described by E.B. Ford in his seminal "Butterflies" (the first New Naturalist, published 1945) "when alarmed the whole brood jerk their heads upwards with a strange unanimity, producing a striking effect." It is a disconcerting sight. You think you are eying a collection of individuals, but this act of collective movement suddenly transforms the mass of caterpillars into a single rather alarming organism, one you might think twice about before tangling with.

It's presumably this protection that gives the species the confidence to disport itself in full view as it reduces your brassica crop to ribbons. But nature has many ways of population control and gregariousness is no defence against disease, quite the reverse. Getting on for 60% of large white caterpillars are killed through infection by viral, bacterial or fungal organisms. Nor does the scent or synchronised movement deter the great variety of insects intent on laying their eggs within the caterpillar. It is reckoned that about 10% of all the myriad insect species are parasitic on other insects for at least part of their lives (parasitoid), and, sure enough, 85% of the remaining caterpillars are said to be accounted for by being eaten alive. The chief agent here is *Cotesia glomerata*, a wasp-waisted creature, technically a braconid fly, about ¼ in./6 mm long. If you grow brassicas (and who doesn't grow at least a few?) you'll probably have seen the larvae – maybe 40 or 50 of them, wrapped in the yellow silk of their cocoons in a cluster around the shrivelled corpse of their erstwhile host. They've eaten all of the fat and now, finally, have consumed its vital organs. Of the dozen or so caterpillars that made it to the tunnel doors a good half turned out to be only empty and now abandoned vehicles for their tiny fellow-travellers. Some of the hosts though get taken short (as it were) so that the corpse and its attendant cocoons can often be seen on the crop itself.

Many chrysalides, a good 90% probably, are taken by birds or small insectivorous mammals (shrews for instance) through the winter, but failing that there is often an even smaller fly, less than half the size of the former, that waits on a metamorphosing caterpillar and lays its eggs through the hardening casing of the chrysalis. It's a chalcid wasp – *Pteromalus* sp. Once my eye fell on a chrysalis just at the moment that a hole not much bigger than a pinprick opened in its side and dozens of tiny metallically sheened insects issued through it like so many Greeks from the Trojan Horse, before flying off into invisibility. This was in the days when we had a couple of acres of brassicas and during the summer, so this would have been the first generation – when the whole process takes place over a few weeks, rather than months. I reckon it to be one of the most remarkable sights of my life.

What the above adds up to is that out of 10,000 Cabbage White caterpillars only 32 achieve adulthood – 0.32%. Now you can see why there are so many of them to begin with.

NATURE NOTE 35

Leeks, from salt to silt

Spring 2018

It's a curious thing that the wild originators of so many of our vegetables are maritime plants. All the leafy brassicas, down to and including kohlrabi, are cultivated races of *Brassica oleracea* – the wild cabbage. Specimens of this can be found all around the coast wherever there is chalk or limestone. I remember them from the quarried limestone cliffs of the Isle of Purbeck, stout, woody at the base, their tough leaves mottled and weathered – indubitably cabbage, but not the sort of cabbage you'd see in a garden – somehow they flourished out of fissures and crannies in the rock. Sea beet (*Beta vulgaris ssp. maritima*), an unprepossessing and sprawling perennial and a common back-of-the-beach plant, is reckoned to be the origin of all the root and leafy beets. Give your beetroots a spray of salted water and you'll see the truth of this in the gloss it brings to their leaves. The cultivated asparagus (*Asparagus officinalis*) is now commoner along our coasts as an alien than the related *A. prostratus* which is native to that habitat. Even the wild carrot (*Daucus carota*), common enough everywhere on alkaline soils, flourishes particularly when within sniffing distance of the sea.

Of the three major Allium crops both onions and leeks occur here only as casual escapees from cultivation. Garlic (*Allium sativum*) however has naturalised itself at two seaside sites – Porth Dinllaen in N.Wales and the Lune estuary near Lancaster. Onions and garlic are believed to be of central Asian origin and were being grown in Egypt around 3000 BCE. As for leeks – we know that the Israelites were eating them before their Exodus from that country sometime towards the end of the second millennium BCE. According to the Revd. Johns (Flowers of the Field, 30th edition, 1902) they were held sacred by the ancient Egyptians, but then – what wasn't? The cultivated leek (*Allium porrum*) has no known wild equivalent but shows an affinity with the wild leek, *A. ampeloprasum*, which is native to all the shores of the Mediterranean and the Black Sea and also to the Atlantic islands (Azores, Madeira etc.). It occurs sparsely in the U.K. on rocky or sandy sites close to the sea, chiefly in the Southwest. I've never seen one. Taking the growing conditions into account, as well as botanical differences between the two species, I imagine its resemblance to a cultivated leek would be slim. We shouldn't assume that salt air is necessarily the reason for this and the other vegetable progenitors' liking for the coast. The equable micro-climate and the alkalinity of sea sand may be alternative factors.

The cultivated leek arrived in our country (if it wasn't already here) during the four centuries of that earlier epoch of globalisation known as the Roman Empire. We know this because archaeology, a discipline which hasn't much to go on but the rubbish that gets left behind, can now deploy micro-analytical techniques that allow the identification of plant traces in cooking pots and latrines. It's remained a staple ever since. In the obscure centuries that followed the dissolution of the Roman world, when missionary saints crossed the Irish Sea on millstones, oak leaves and other improbable craft, we catch a glimpse in Celtic poetry of the hermit in his monkish cell and of the leeks that sustained him growing in his garden plot. Then there's the Welsh king Cadwallader battling against the Saxons in 640 CE. His warriors wore leeks gathered from a nearby garden in their hats as distinguishing marks. They triumphed, and hence the emblematic Welsh leek. On a more workaday level the late fourteenth century Piers Plowman grew leeks in his croft, which provided him also with peas, beans, parsley and "chiboles [small onions] and chervils and cherries, half-red".

Leeks circle the year, the seed of the new crop going in while the last of the old is harvested. The leek is not a plant in a hurry. From sowing it takes twice the time that brassicas require to make plantable size, and if you want ripe seed you must wait until near the end of the second summer when it's hazarded all on one attenuated central stem and the eventually opening pom-pom at its tip. It has a single-mindedness of form and none of the bustling growth of the cabbage-kind. It simply doubles its first thin monocotyledon into an elegantly symmetrical column until sooner or later (sometimes too soon for the grower if things are not to its liking) sending up the single flower stalk that echoes its first shoot from the seed. The leaves slough away until there is little left but that disproportionate flower-head so beloved by the bumble bees, and then finally its however many thousands of seeds.

We planted 50,000 leek plants in our first summer here. All but a few thousand were destroyed by wireworm in a drought (we had only our tears for irrigation) and from the survivors we harvested less than 500 pounds. We persevered. We put up our first polytunnel so as to be able to raise our own plants, easily done under cover. Once we'd got the hang of things we were harvesting sometimes 1,000 pounds a week through the winter – all in ten pound nets! Grown on ridges as an aid to drainage and harvesting they flourished mightily on their five-inch spacing so that it was generally a struggle to make up nets that contained even twenty leeks. Both they and onions responded well to our silty loam and – if you'll excuse the pun – leeks went from being a source of tears to becoming our flagship crop. I enjoyed growing them and despite outbreaks of rust and white blotch and the days when frost turned them into columns of ice and the pain in the fingers became almost insupportable, I usually enjoyed harvesting them too. It's the single activity that I miss most now that I'm no longer a commercial grower. In company the easy rhythm of the work allowed conversation to flow. Alone, the task left mind and spirit free to take in the comfortable surroundings of field and hedge and more distant wood, with the flight and song of birds and all else that the world freely offers. Our work binds us to the land, but there is a freedom there too which is infinitely precious.

NATURE NOTE 36

The opposite wood

Summer 2018

Gazing out over an English landscape of fields and woods it seems reasonable to suppose that change is always towards greater exploitation, less wildness and in particular less trees. For much of the country this is probably the case, but in our own little valley and in the greater valley into which it feeds the opposite applies. Here the retreat of farming due to the agricultural depression of the 1930s was not entirely stalled by the post-war policy of financial support for production. It was then encouraged again by golden handshakes aimed at "rationalising" milk production and by the inability of somewhat marginal land in smallish parcels to keep up with the industrialisation of arable crops. In the three and a half decades since we came here the sound of the forage harvester and combine has been largely replaced by that of the topper and ride-on mower.

There is a wood on the steep slope opposite to us across the little valley of the brook. It forms the backdrop to our life here and I look into it through the window as I write. Its changing countenance through the seasons – sunlight on the bare stems of winter, the gorgeous vibrancy of its spring awakening, the heavier green of summer maturity, the wind-borne rain that can be traced against its darkening autumn colours and the curious vapours that may unpredictably rise from it and hang there during steady rain on a still day – these are familiar, appreciated but unremarkable. So too the flight of the ravens and buzzards that nest on its upper edge, often enough locked in their immemorial squabblings as they rise into the sky above.

There is a cut through it, thirty yards wide, that runs first obliquely from its lower edge and then straight up to breast the skyline. This accommodates the stout wooden poles of a high-voltage line supplying electricity to our neighbouring village, installed in the 1930s. Due to the convexity of the slope two of these poles are outlined at the summit, their cross-arms clear against the sky. One more would make a Golgotha. Every so many years a chainsaw gang works its way up through, clearing back encroaching growth. Looking at this gash there seemed something brutal in it and I remarked as much to John Clarke, the dear and now departed contractor who gave us so much help with his tractor and digger, between mechanical breakdowns, when we were first here. 'Oh no' he said, 'the powerline was there before the wood'. This seemed to be turning things upside-down, but looking more closely across the intervening few hundred yards it was possible to see that most of its trees were indeed young ash and sycamore, poles rather than stout trunks. Later on when we penetrated it (getting on for a mile away along the public road) we could see that it was only on the old hedgebanks, now invisible except at close quarters, that there were mature trees. The unprepossessing ground flora of ivy and hart's tongue fern with here and there clumps of wispy etiolated brambles is a classic of secondary woodland. The banks are shown on the early twentieth century OS map enclosing a few fields of 2 or 3 acres each. Only at one end was there a small area of established woodland.

The slope is formed where the cleft of the brook descending from the high land above opens into the greater valley of the river. On the right bank the heights run away southward in an escarpment with some (by local standards) reasonably level ground at its foot, but on the left bank and northern side they extend into a gradually diminishing spur. With the help of a fault in the strata of the underlying rock the brook has gouged away at the side of this spur so that it rises 250 feet in 600 (measured on the 1:2,500 map) and in places approaches a scree slope, stones shifting and slipping as you tread. It seems extraordinary that this ground was ever cleared and enclosed. The only possible thing in its favour is that it slopes south at perhaps an optimum angle for absorbing the sun's rays, and John remembered early potatoes being grown on a less steep patch of it.

Such evidence of land hunger and past use is far from being unique to our valley. The hopeful nibbling away at land that has since been abandoned can be seen, for instance, in the form of mediaeval lynchets (terraces resulting from cultivation) on Dartmoor moorland and on some of the steeper slopes of the Wessex downs – ground that no farmer, however rapacious, would look at now. The threat of dearth and starvation brought about a high tide of settlement and exploitation in the thirteenth and fourteenth centuries which only ebbed when the Black Death dramatically reduced the number of mouths to be fed, but there are nineteenth century farms too formed out of marginal land and soon taken back by the heath and moor from which they were wrested. Elsewhere some of the most extensive evidence of aborted land hunger can be found in North America. There's woodland in Wisconsin that, despite looking to the unfamiliar eye as if it had been there for ever, has grown back over land which had been cleared for agriculture within the previous century. In New England there are an estimated 100,000 miles of stone walls, enough to circle the world four times, now under dense tree cover. The fields they enclosed and the homesteads which they supported for a while they supported are as if they had never been.

To anyone with an empathy for the extraordinary labour demanded by this clearance and boundary-building it is a sad, even a painful sight. But it is a hopeful sight too. I can't believe there can be any more appropriate covering for the slope opposite than the trees that now grace it. Trees will grow practically anywhere that the climate allows. They don't need to be planted or cared for. They just need to be left alone (by humans, farm livestock and now by deer) for a few years of establishment. Whatever mess we make trees will redeem it and outlast us as they do so.

NATURE NOTE 37

Water's ways

Autumn 2018

When rain falls on dry ground the moisture moves into the soil along a broad front, each water-holding pore becoming filled before releasing the excess to pores beneath it. Deeply cracked soils, those with a high clay content, may exceptionally lose water to drainage before the entire profile is wetted to capacity, but generally if you know both how much rain has fallen and the water-holding potential of your soil you can have a fair idea of how far that rainfall will have penetrated. Available water capacity (AWC) varies from less than 1½ inches per foot in sands and loamy sands to over 2½ inches in peats, silt loams and very fine sandy loams. Surprisingly (perhaps) the AWC of clay soils lies between these two extremes. This is the available capacity – unless dried in an oven there is always some water in soil but the residue is so tightly held that plant roots accustomed to humid conditions cannot access enough of it to maintain growth, despite exerting an extractive force of 10 atmospheres, equivalent to a column of water 300 feet high. At this point the plant wilts – the leaf stomata closing up to prevent further transpiration. The roots can struggle on beyond this, exerting a suction of 20 to 30 atmospheres to keep the plant just alive.

Soil of course is not a solid – about half of it is space filled with air and water. The particles and aggregates of particles that constitute its mineral composition provide a surface area that in an acre of fertile topsoil may equal 5,000 square miles – more than, for instance, the land area of the counties of Devon and Cornwall combined. To help shed light on the unfamiliar reality of bigness within smallness that results, Sir E. John Russell explains (in The World of the Soil) that a crevice 25 microns (1/1000 inch) across (far beyond human visibility) is as spacious in comparison to a molecule of oxygen as a valley 120 miles wide is to us. Water molecules are smaller again, not much more than half those of oxygen. Water is held in pore spaces that are 30 microns or less across. Smaller than about 7 microns and the water becomes impossible to access – it can only be removed by heating the soil in an oven. Larger pores will fill with water from heavy rainfall, but then drain to admit the air that is as vital to plants as the water.

Evaporation during the growing season is rapid from a damp surface but soon inhibited by a shallow layer of dry earth and pretty much ceases once the top 2 inches have dried out. If you need the soil to be dry enough for deep cultivation a previous crop or green manure is advisable, even in a drought. On the other hand brassica and leek transplants – peg plants whose roots can be got in deep, not the modules that we now all seem forced to use – can be established if not too much damp earth has been exposed and thus allowed to dry out during previous cultivations. We used to worry much in a dry time, especially as at first we had no water for irrigation, and then just a little, but generally we ended up better off than in a wet season. It was something to see those transplanted six-leaf cabbages crisp up and apparently die before, within a few days, sending up a green shoot and from that grow on to gainful maturity. But this is on a quite deep silt-loam over a gritty clay subsoil. It might have been a different story on a light loam over shillet.

I never wanted to see a repeat of the drought of 1976 and, despite the fearsome dryness of this summer, seem to have been spared it again. Some rainfall in July – ½ inch early in the month and again at the end – served to brighten the veg garden a little but made no difference to the pastures and the heat continued. In the last few days (I write mid-August) we've had what's amounted to an inch or so and cooler weather with it. We're holding the cattle on yet another bale of silage in the hopes that this rainfall will at last bring life back to the grass, which so far has had only a few weeks of growing conditions since last November. The frustration with grass ground is that the surface mat of vegetative material holds a good part of the precipitation which then evaporates off rather than penetrating the soil, whereas the open nature of cultivated vegetable ground allows most of it to find its way downward – and wetting even of only the topmost part of the root-zone has an obviously positive effect.

What made 1976 so severe was that 1975 had been one of the driest years ever recorded so that there was little reserve of moisture to draw on. Long gone now, it is memorable mainly as the summer that induced men to wear shorts (which only eccentrics did beforehand) and which saw tables and chairs appear on the pavement outside cafes and bars where they'd never been seen before. We'd spent the previous two years in the north of Scotland where rainfall remained at least adequate, coming back down to the south of England in late June when already the heat was exceptional and the land burning up. We went to Ontario for six weeks, all those lakes, no lack of rain and no idea of what was happening at home (in the age before mobile phones and internet). Penetrating the hazy cloud as the plane came down to Heathrow I thought for a moment that we'd taken a wrong turning. Surely this was Algeciras, if not Khartoum. At my mother's in Dorset the air was thick with smoke from heathland burning underground. In Devon, tramping the stifling lanes in search of somewhere for us to live while I went to agricultural college, I came across a mole apparently similarly engaged, but with no hope of penetrating the rock-hard ground. The Exe was barely flowing over its weirs. When with the August bank holiday the rain came at last, like a benediction, the break-up of the river's rafts of duckweed and scum was a sight to see.

A bit about ivy

Winter 2018

There were a few mornings of moderate frost at the end of October and into November, followed by floods of rain and strong winds. The frost disrobed our three walnut trees overnight so that they seemed shockingly naked, the rags of their clothing covering the ground beneath. Wind and frost combined soon dislodged most of the leaves on everything else. The rain was enough to get the springs flowing again and after months of quiet the air was full of the sound of running water. One way and another it felt that summer had been cut and stripped and washed away, no trace remaining.

Then on the 11th after 4½ inches of rain in a week the morning came clear. The sunlight at noon slanting over the thatch of the little bank-barn fell directly on the opposite wall of the yard a short distance away. This wall, un-mortared and substantial, is for part of its length surmounted by a billowing crest of woody ivy four to five feet high. In such places, free of shade, ivy blooms generously from late September on and provides the last floriferous feast of the year. By mid-November many of the former flower clusters are swelling into berries but there are still plenty of open flowers rich in nectar. The wall hummed with airborne life, backgrounded by the laundered blue of the sky and the shining glossiness of the ivy leaves. At the flowers were a surprising number of wasps, numerous bees with their hind legs bundled in golden pollen, a hoverfly or two and the buzz of other more nondescript flies. Here was a tableau of summer while just along the wall, where it merges into an earthen bank, autumn held sway with hips and berries and hazel stems festooned with the dead bines of wild hops.

Ivy feeds birds too. Soon after Christmas the last yellow trace of the flowers is gone and the berries fill out from dull green to matt black. Their fat content makes them nearly as full of calories (weight for weight) as a Mars bar, but it's not until the hungry gap of early spring that the birds appear to relish them much. I've seen it suggested that this is a conscious restraint, but birds are not known for their restraint – it's not a quality that would benefit their ungoverned lives. This year the hollies have displayed an abundant crop of berries, bunched like grapes. The birds have had most of them already – once ripe enough to eat why wait? It would only be for some other bird to snaffle them. Probably the bitterness of ivy berries, which makes them unpalatable to humans (fortunately, as they are mildly poisonous to us), is worked on by time and winter weather so that they come ready for the birds just when there is little else about and they are most needed. We might see the workings of Providence in this matter but not, I'm pretty sure, any avian self-regulation.

The window by our kitchen sink gives a good view of the wall and its ivy a few paces distant. In fact it's most of what we see whenever we run a tap or do the washing up, as it blocks the further and rather more romantic view westward. On this account and also because the thick mass of wind-resisting stems and probing roots ought to threaten the integrity of the wall, and again for what it says about a tidy farm (some hope!), we might have cut it out.

On the other hand the ivy was there long before us. We cleared the yard, which had been reduced to a path the width of a bullock, but we never wanted to do more violence than necessary to the habitat that the years of the farm's abandonment had produced. So the ivy remained. It's mainly blackbirds and thrushes that eat the berries, and these days wood pigeons too, alas. They give every sign of enjoying them when the time comes. The pigeons loll about on top of the supportive vegetation while they pick contentedly at those fruits within easy reach. The fair-sized pips pass through the gut primed for germination. Before they do so they're easy to spot on account of the peculiar lilac shade of the aril, the seed's outer coat, which delivers a little shock to the sight when the eye lights on them, the colour being so foreign to the place and season.

There's more to ivy than autumn flowering and timely fruit. These are the visible face of a microcosm. The inside of an ivy tod, to give the woody clumps their proper name, is a disordered tangle of stiff and mostly naked stems which at any season offers security from oversight, and shelter from sun and rain and storm alike. The mix of live and decaying stems and their flaky bark is a refuge for over-wintering and hibernating insects. It's also impenetrable to hawks and at least tricky for cats, so it's a refuge for the small birds too. In summer the leaves feeds caterpillars, notably that of the Holly Blue butterfly, jewel-like when it first emerges and not shy in showing itself off.

Ivy is remarkable for its own metamorphosis. The young leaves are three or five lobed, dark with sometimes a hint of purple, strongly veined and borne on brittle stems. It's only when these stems have become woody and sinuous and climbed into the light, a few feet in the case of our wall or many when reaching the extremity of a good-sized tree, that the leaves lighten in colour and simplify into an un-indented and rounded form that bears little relation to what came before. Now at last can the plant produce flowers and fruit. All plants seek light, but with ivy this governing principle reaches its fullest expression.

NATURE NOTE 39

Stones

Spring 2019

It wasn't something I'd ever done before while at work, but asked to carry on picking up stones after tea I thought a wee smoke might help the task along. And so it did. Picking up stones and digging out rocks (they grow, year by year) from the arable was something we did a good bit of on that Ross-shire farm in springtime. It had its moments as when we found a stubbornly large boulder for Derek to tussle with from his seat on the back-actor, mounted on an elderly International kept for this purpose. Lifting the smaller ones with a four-pronged graip ('fork' to Southerners) and pitching them into the trailer on interminable bouts back and forth across the thirty acre fields had less to recommend it. But that evening in the clear light of a northern spring, with time slowed down a bit and the veil parted, I could appreciate that each stone had its own individuality. Born of ancient Highland rock of different shades and textures, gouged out, ground and shuffled about by the advance and retreat of successive glaciations, each piece was different in form and heft, veined or flecked, often crystalline and not uncommonly studded with tiny garnets. Stone, I reflected, has its own life – if lived at a very different rhythm to our own.

Back in the Westcountry, where there never was glaciation and where the rock beneath is the likely parent to the soil above, we were six years on a farm underlain by shillet. Shillet is the Westcountry term for a shale rock formed out of sedimentary mud by the ponderous weight of geological time. Grey or pale brown, it's easily fractured, particularly in the horizontal plane, and gives rise to smallish flat stones, so that there was seldom anything that needed picking up on that farm. As shillety soils go this wasn't a bad one – a decent loam, grade 2 where level enough – but even so the saying was that the parish needed a shower of rain and a shower of shit once a week. For shillet is free draining and provides only the stingiest reserve of moisture in the subsoil. It's true that it's handy to have on a farm for its value as hardcore, but it's dull stuff – as I got to know well when digging post holes. Once not too far below the surface the way down was only with the point of the iron bar, flake by effortful flake.

Shaking the dust of chemical agriculture from our feet we found ourselves in a different valley with a very different geology. What underlies at least part of this farm eventually came to light when drilling 120 feet down to make a borehole and turned out to be a soft black shale. Otherwise even quite deep excavations revealed only a slow gradation from topsoil through barely differentiated subsoil to a base of gritty clay sprinkled with unremarkable stones.

We do have rocks, but they occur from the top down. Long ago cleared from the surface of the land into hedgebanks and bits of dry-stone walling, there yet remained plenty of just submerged boulders to confound the mechanised vegetable grower. From the tractor seat I thought of them as sharks, lying in wait unseen beneath the surface. Some were small enough to be no more than irritant, others capable of mangling a plough beyond use. One answer was not to plough, another – after too many trips to the blacksmith – was the purchase of a ridger with sprung-mounted bodies. Another again, when we started to level the site for our shed and found lurking there a rock that the JCB couldn't even induce to so much as twitch, was to move the shed. Two were standing stones, for cattle to rub on probably. They are still standing. Many we did dig out. The biggest were saved up for the digger's visits, others made to yield to tractor and chain. They sit in odd corners gathering lichen and moss, shapely and comfortable in the dignity of their immobility.

These stones are not native to the ground, but they haven't come far. Discounting the theory of a passing geology student that people had put them there "because they were stupid" I imagine they arrived by gravity. A little way back from the sort of terrace on which this farm sits there rises the escarpment of the granite plateau above. When the granite originally intruded from the magma into the earth's crust, the intense heat baked the adjacent rock, altering its nature and driving veins of minerals into its fractures. Hence the narrow band of the mineral-rich metamorphic aureole that surrounds the granite. The local version of the resulting stone is generally known as blue elvan, though the older inhabitants call it simply 'woodstone', in distinction to the 'moorstone' (i.e. granite) on the plateau beyond. Hornfels diabase seems to be the technical term, though it is sometimes called basalt. It's what the escarpment is composed of, never far from the surface and outcropping here and there. Excessively hard, resistant to fracturing and damnably heavy, it's not easy stuff to deal with. Freshly broken it's an intense steely blue in colour fading on long exposure to a blue-tinted grey, though individual rocks often acquire a dull brown crust through some process of weathering.

A substantial tor, not of Dartmoor granite but of this elvan, formerly stood six hundred feet above us on the edge of the plateau. Scat Tor it was called, or Scatter Rock. No doubt just as the tors of the moor proper were once mightier piles, which in the aeons of decay since their original exposure have littered acres of ground beneath them with tumbled stone, so too with this one. It's hard to visualise now – you'd think the stone would have stopped well short – but you have to think of a time when a mighty pile of rock towered above and deeply frozen ground beneath offered no impediment to the downward movement of its parts. Described by a now departed village historian as, in its latter days, the two hunched shoulders of a giant flanking a space between that made a nice den for children, the tor is long gone. It was blasted into Scatter Rock Macadams' hill top quarry, now a two acre lake, not long before operations ceased around 1950. Crushed and sent out for roadstone – that you might think was truly stupid.

The process by which those rocks came to litter our ground is 'solifluction'. Away from the ice sheets that covered more northern parts the conditions here would have resembled arctic Canada or Spitzbergen today. The stones worked their way downslope lubricated by the brief layer of summer mud overlaying the permafrosted ground beneath. Apparently a slope of as little as 5° is enough to achieve this movement.

NATURE NOTE 40

The turtle's voice

Summer 2019

The heat of Easter brought on a flurry of insect activity that in a few days had seemed to outdo the total insect presence of last summer. This is not to say a lot, last year having been so lacking in entomological life. All the same it was cheering that by the end of the month there were more ladybirds, together with more aphids for them to feed on, than in the whole of last year's growing season. So too a variety of butterflies, some hoverflies, numerous assorted bees (bumble and otherwise) and an extraordinary profusion of oil beetles (see note 20) lumbering about in pursuit of their curious lifecycle. The winter must have been kind to them all. Not so the birds, whose decline appears to continue apace.

I've been interested to read the wonderfully named Isabella Tree's 'Wilding'. Though I don't altogether go along with all of its message it is a powerful statement, well-written and forthright, of the mess we have made of the land. The chapter on soil life ought to be especially enlightening for anyone ignorant of what organic farming is about. That it's there at all is surely a signifier of one of the great successes of the organic movement – the recognition (not before time, and not least by Defra) of the centrality of the soil to life on earth. It struck a chord that a turtle dove features on the book's cover and that the author places the return of these birds at the centre of the success of the Knepp rewilding project. I can't reasonably expect to hear turtle doves in our locality because even when there was still mixed farming here and arable weed seeds to feed them, it would have been on the edge of their range as summer migrants. But as a teenager I lived on the further side of Dorset and one memory of their presence there has stayed with me for fifty years.

It was sometime after A levels, high summer. I was returning from a walk up into the chalk country north of our home. I'd slept the night in a patch of unimproved downland, an area of bushes and flanking grassland, where the line of a Roman road was marked by a hollow way as it climbed the steep bank of a little clear-watered valley. At dusk and the skylarks falling silent I'd laid out my bedroll on the flower-spangled turf. Lying comfortably there under the overarching sky of that open land and with the darkness quietly gathering, the grasses and herbage that fringed my resting place came to reveal another aspect of their life, one that I'd never witnessed before – and only seldom and then partially since. Each stem and blade of vegetation became lit with an etheric otherness, an aura if you will, each one sheaved in a silvery light, that shimmered and danced and extended into the stillness of the night. It was not (obviously) a physical manifestation, like the light projected by a glow-worm or the gorgeous phosphorescence dripping from the lift of oars in calm salt water, but seemingly an emanation of the vital spirit of the plant. And the light was complete in itself, contained, arising from all the myriad plants within my view but not in any way illuminating the night around. No doubt Rudolf Steiner could have explained all this. For myself I'll just say that I have never since wanted to make any distinction between the quality of life, or indeed consciousness, in plants on one hand and animals on the other.

What of the turtle doves you ask? I'm coming to them! Whether the auras faded or I just fell asleep as they shimmered I cannot now remember, but after a few short hours of good repose the faintest touch of dawn had the larks again mounting the heavens, seeking the sunlight long before it touched the ground beneath. No doubt I had some sort of breakfast before setting off homeward. Before noon I was off the chalk and on a grassy hilltop in the belt of clay-with-flints that lay between it and the last few miles of heathland. Here I sat at the base of a wide-spreading tree to take some refreshment, replete with the loveliness of the night before, the beauty of the limpid day that had followed it and the pleasant feeling of energy well spent. It was then that I became conscious of the purring of a pair of turtle doves close by. The bird's specific Latin name, turtur, well illustrates the sound, though for verisimilitude you need to add two or three 'R's to each syllable. It is, as Isabella Tree tells us, as close to a lullaby as the world of birds gets. Low-pitched, almost a moan, it is a sound so soothing, so peaceable that it becomes like running water – the harder to move away from the longer you are within the hearing of it. Though I love birdsong I don't have a good ear for distinguishing or remembering it, but the gentle crooning of those turtle doves and the sensations it induced remain with me, undimmed I feel, after fifty years.

Turtle doves are farmland birds. In truth most of the land birds that inhabit the UK are farmland or (at any rate) garden birds, seeing as almost all of the country is farmed or gardened. There's a grid pattern of fields still extant in a part of Essex that is slighted by a Roman road, fields in Cornwall that are believed to be Bronze Age in origin, and so on. Farming is not new. Land clearance continued for maybe five thousand years, reaching a mediaeval high tide that ebbed only with the Black Death in the 14th century. Birdlife coped with these changes, some species declining, others increasing, but a balance and seemingly an abundance always being maintained. I don't believe that our problem of lately losing the ability to live in a mutually beneficial way with birds, and with the rest of wildlife too, is answered by making over large areas of agricultural ground, even of the poorer sort, to wildlife reserves while the more productive ground remains an agricultural desert. One hundred percent organic, and a prosperous peasantry on the ground – that's the way to do it!

Rudolf Steiner (1861 – 1925) – founder of the Biodynamic school of organic agriculture. Its philosophy and methods "recognize the importance of the healthy interplay of cosmic and earthly influences".

NATURE NOTE 41

Nuts!

Autumn 2019

It's mid-July when the squirrels start on the hazel nuts, ripping off whole clusters, biting through the tip of each developing nut and nibbling out the tiny infant kernel, before discarding the ruined bundle to litter the ground beneath. What a waste! Can't they wait a month until there's something there worth having, or better still until late September when the mature nut is packed with 16% protein and 60% fat? It's what the dormouse needs to fatten up and keep itself in good fettle through a sleep of six months. Enough also to sustain the mice and voles, and the nuthatches and tomtits, that 150 years ago never had to cope with the demanding competition of grey squirrels. In a well-ordered countryside there'd be a bounty on the creatures, or better, a population of martens which, now pushed to the margins, are skilled killers of squirrels. Either would leave more nuts for the rest of creation as well as reduce the damage done to timber and nestlings.

But the temptation to have a go at them when still green and only half-hard is understandable, given the great effort that rodents and birds have to put into penetrating the fully ripened shell. Of all the rodents the squirrel is best equipped for this challenge – holding the nut with its forefeet and gnawing a groove at the apex until a small hole is made. It then inserts its lower incisors and uses them as a crowbar to part the shell. Other, smaller animals with smaller paws and lesser teeth need to work longer and with greater determination before reaching any consummation. When we lived in our caravans in the lee of a tall hedge of hazel we were sometimes entertained (and sometimes irritated) by the sound of small rodents rolling nuts around above us as they grappled with the task in the narrow space between ceiling and outer skin. Comparing the weight of a hazel kernel (even if the wild is only half that of the cultivated, which come at about 1.4 g) to the weight of a mature female field mouse (at 14 g) you can appreciate that there is every incentive to keep at it.

The squirrel (650 g) is a vigorous and apparently highly-strung animal with a liking for chucking things around. When we grew parsnips on a fair scale they took a fancy to them and would carry reject roots (unlike badgers they didn't dig for them) up into nearby trees and shower the surrounding ground with coarse flakes of peel and other bits that they discarded from the feast.

Now in mid-August, when the developing kernels have expanded to half fill the shells, the litter beneath the hazels increases. Some of the shells have been pecked by birds but it is still mostly the work of squirrels, several of the nuts now expertly parted into neat empty halves but plenty just dropped and forgotten about. Later on they will put some effort into stowing away part of the harvest, usually in a random way by pressing individual nuts into soft ground but sometimes in small caches. They've been known, from tracks formed in snow, to at least occasionally set a definite course for these. Even if the act is random, what looks like a good place to bury nuts will also look like a good place to find them – so the scheme works well enough. It also provides hope for scavengers and, in those that get missed, for the tree's posterity. Many other rodents store autumn food, sometimes – as dormice and water voles – within reach, never mind memory, in their winter nests. The most remarkable example of this propensity that I witnessed involved a well-grown autumn cauliflower which, when cut, revealed the score or more of acorns that had been rammed into the narrow spaces between the bases of the leaf-stalks.

The hazel is a long-lived tree, constantly regenerating from its stool and (here at least) quite capable of perpetuating itself through the odd nut that happens to get buried in some suitable spot. What route the rest of its annual crop of fruits and leaves takes back to the soil's carbon store is of no moment, it will be feeding organisms of one sort or the other on the way there. The natural world is infinitely fecund. It has its seasons of course and its generosity is not constant. Nor is its fecundity without fickleness, as growers well know. But when there's food to be had, there's food to be had and it's a free-for-all. Winter storage is exceptional and the squirrels' attempts at it are far outweighed by their spendthrift nature. Carefulness seldom come into it. Take what's best now, don't leave the choicest bits to later – something else will have them! See the scattering of half-chewed flesh and discarded pips left by birds beneath a fruiting bush in autumn, the one bite taken out of each swede by a passing deer, the peas or sweetcorn trashed overnight by a gang of badgers. 'Waste' is a loaded word to apply to this behaviour and surely not one that nature recognises. What one organism leaves behind is another's sustenance. Everything is sooner or later on the way back to what it was, and every route there (unless fire comes into it) involves a good deal of digestion, a process that reaches its end in the soil.

Waste is something that we humans create, largely because much of what we leave behind is now outside the cycle of digestion, being either toxic or resistant to decay, or both. Our carelessness is seldom benign. Back in the spring and needing material for a deer fence I contacted S---- at the sawmill up the road to be told – 'I don't know Tim, I'll get it somewhere but I'm short of wood.' When I saw him later, with long strainers that he'd had to buy-in from elsewhere, he explained the problem. A fleet of articulated lorries takes full loads of subsidised timber week after week from Devon to Kent to burn in a power station. There's no such profit in local trade. Who cares? Carbon to carbon dioxide, a sawyer short of wood – that's nuts for you!

NATURE NOTE 42

Winged life

Winter 2019

A source of entertainment over some bright days around the middle of September was the sap run on a decayed Turkey Oak that stands, for now, on our boundary hedgebank. In truth more an ooze than a 'run', the sap glints darkly within fissures in the bark over a wide area of the trunk from the top of the buttresses to well above head height. The viscous substance that comes to the surface settles out into an unappetising grey goo which was proving to be a magnet to late summer red admiral butterflies. These arrived on shafts of sunlight, a constant trickle of single individuals and twos and threes. They were all newly minted, freshly emerged from some not too distant nettle bed. At any one time there were half a dozen or so settled on the trunk or flying around it, with countless others at rest on nearby vegetation – these presumably sated and digesting whatever magic the sap provides. It must be sugar of course, though that's not what you'd expect from a Turkey oak, apparently at least as well endowed with tannins as the native kind. Red admirals are suckers (literally) for sweetness. The flowers of high summer which provide them with nectar have faded, those of the ivy are yet to come. September brings the juice of fallen fruits – they love pears and plums especially – and this sap is a welcome bonus.

It wasn't just red admirals – there were hornets too, some feeding on the sap, others flying about close by. Like wasps, they want only sweetness at the end of summer when the work of the nest is done and the last of the queen's output has been left to starve. Now and again one or two of these hornets would fly at a red admiral and harass it into flight. Earlier in the summer butterflies are caught, carried back to the hornet nest and chewed up to be fed to the larvae, but these attacks seemed to be more an enforcement of the pecking order than naked aggression. In each case peace was soon restored and the bullied butterfly back to feeding elsewhere on the trunk.

Large flying insects like hornets and butterflies are easy enough to spot on any bright day during their season, but though we know (if we think about it) that there must be winged life there, because we see the martins and swallows hawking close overhead for their prey, actually seeing what it is that they are flying after is seldom possible. Once, just once, I had a small moth snatched from sight no more than two or three feet in front of my nose. The snap of the swallow's beak was audible. Occasionally though, towards the end of a fine summer or autumn's day when the air is still and you happen to be looking out from the right sort of shadow, and the view is back-grounded by something light-absorbing, the sun's beam comes at just such an angle that it illuminates an astonishing mass of flying life, a host of more or less diaphanous forms otherwise unnoticed but now seen to fill the air above.

Warm days in late summer and autumn often induce large aggregations of certain winged insects. One species that can form clouds sometimes mistaken for smoke is a type of frit fly, *Thaumatomyia notata*, which is distinctively gold and brown and no more than 3 mm. long. We get these in the loft of our shed, swarming against the skylights. As the roof is at best only a little above head height you want to keep your mouth shut while you're up there. This and other species of similarly diminutive flies sometimes congregate in houses and may use the same spot over a succession of a dozen years or more. Very likely they have some refined sense, we might call it smell, that guides each new generation to the spot as surely as sight guides us. My own now distant sight of such a congregation involved a different, darker species and July, rather than late summer. Going up to bed in a rambling old farmhouse I found part of the wall and ceiling of the room coated with thousands of these tiny flies, like a map, a dark continent marked out against the white of the paintwork. Despite my switching on the light they all remained quiet and unmoving. When I woke every last one of them had left, presumably through the open window. I thought this a singular achievement, given the mystery that glass presents to most flying life.

More annoying are the cluster flies, *Pollenia rudis*. A bit bigger, blacker and glossier than a housefly and with wings that when at rest overlap like scissors, these are driven by cooler weather at summer's end to seek out west-facing walls where they can bask in afternoon sunlight. They find cracks around the window casements and eaves into which they cram themselves overnight, along with as many of their fellows as is feasible. If you open a window after dark many will come through it and spend an irritating few hours flying noisily around the light fittings in an aimless manner. They don't necessarily fall silent once the light is off. In such cases you may be glad that some cobwebs have been overlooked. Opening the window in the morning sets them going and a few will come in, but these can be easily ushered out into the daylight. Eventually all cluster flies seek undisturbed quarters for hibernation, more likely in lofts and outhouses than behind a picture in your bedroom, though that is possible. I once had the job of stripping corrugated iron which had been laid on top of some sort of boarding. The voids were crammed with the creatures, each pressed tight against its neighbours. The sudden access of daylight and winter air and the consequent desperate huddling or attempted flight of the semi-comatose flies was a pathetic spectacle, but it's hard to have much care for that when you know that their larvae are parasitic on earthworms.

Small lens – big world

Spring 2020

The sallows begin to show the silvery-white buds of their catkins, by which we know them as pussy willows. At least the males do, the females of the species for now remain furled, though in this soggy season the fat pendant raindrops that catch the light as they cling to the twigs are an almost substitute.

We all love these pussy catkins, even if it's sometimes hard to love the tree that produces them. For one thing they're a cheerful sign of the new season. For another in their velvet softness they're wonderfully sensual – for a tree, especially perhaps for a sinewy willow with its dull grey bark and humdrum foliage. Touch them against your cheek to feel their delicacy – surely they have more of the animal than the vegetable to them? But to the sight (if it's cats we're thinking of) I'd say they most resemble voles, or maybe angora rabbits, as in their little groups they crouch along the stems all plump and featureless. To really appreciate this likeness it helps to look at them through a hand lens. At ten times magnification you can see how extraordinarily deep and lustrous is the fur that covers them, a snug pelt to keep warmth in and wind out. You can appreciate how much energy the tree has put into the comfort of its developing flowers.

My 10X/23 mm hand lens is a quite recent acquisition and because with age my sight is not so keen as it was it's perhaps a timely one, though if I'd thought to get one long ago I might have made a botanist by now. I wouldn't claim that it's an essential tool for a grower but it's an entertaining one. There's almost nothing in nature that is not made more interesting or more comprehensible and often more wonderful by a magnification of vision, and it allows some fleeting access to an otherwise impenetrable world. I say 'fleeting' because actually getting an object in focus and keeping it there is a fine thing. You hold the lens close to your eye, steadying the hand against the side of your face, and then either move the object towards it or move it (and your head) towards the object. The focal distance is a matter of inches, and then too it helps if you can avoid casting a shadow on the object in view. Obviously this mode of operation suits stationary targets more than moving ones, though the other day I did get a good look at a tiny speck of life that was travelling across our kitchen table. I was pleased to discover that seen times-ten it was a good-sized and very shiny and elegant spider. What it saw of me, or of anything else, I have no idea, but it felt like something that I could see it at its level. And that's the thing – the lens gives you some approximation of an insect-sized view, if very likely not the view experienced by an actual insect.

Flowers are great of course, especially little ones, their beauty of form and colour and the detail of their structure is often exquisite and sometimes breath-taking. When it comes to weeds the speedwell may take the crown. And then there's the greenish growths that we generally don't pay much attention to – the getting on for two hundred native grasses and all those funky sedges. There's more going on there than meets the eye! Lichens too are worth a look in all their otherworldly variety – grey-green webs of the foliose kinds hanging from branches, coloured crusty ones on stone and wood, and the shrubby sorts often enough with short stems terminating in little cups. One of these, the red crest lichen, decorates the aging thatch of our barn. Carried a little aloft on irregular silvery stems are the deep red fruiting bodies which give it its name. On close inspection these turn out to be lumpy globular structures that look as if they might be made out of baked playdough and seem quite unlike anything else in nature.

Inanimate objects have their treasures too. I picked up a stone on a beach, a flat disc about the size of a fifty pence piece. Its background colour is a whitish buff, paler than the sand on which it lay. What caught my eye was the glinting of the many tiny mirror-like crystals that stud it, along with a good speckling of others, some black and some white. Viewed through the lens and moving it this way and that the mirrors blink on and off as they catch the light. There are scores of them in the stone's small compass. It might be the surface of some distant planet or a drifting asteroid. And then a single crystal, half an inch in length, came to hand last summer when I was weeding in the vegetable plot. It was grubby and chipped but I thought it more than the bit of glass or even plastic that someone here suggested it might be. Cleaned up and viewed through the lens its smoky transparency and the precision of its faceting confirmed its gem-like nature.

Soil too. It's not exciting in an aesthetic way, but it is at least interesting to get a bit further into the element that's so crucial to the lives of us growers. Put a crumb on some hard surface where you can get a steady look at it. Break it down with a little pressure from your fingertip. Don't worry about the dirt beneath your fingernails (though you can get a good view of that too) but see the mineral matter, the 95% of it, begin to disintegrate into its constituent parts. You'll make out the individual grains of sand and some of the silt, these no more than a fine dust. You won't see the individual clay particles of which there are about five hundred to a millimetre, nor the 5% of amorphous organic matter – for these you'd need a powerful microscope. You do though get some sense of the necessity of both in binding together the randomly shapeless bits that you can see into the structure of soil crumb that make life possible.

Whatever next?

Summer 2020

The rain gauge worked hard for six months and then took a much-needed rest. It wasn't long before the desiccating easterlies turned all that rain to memory. Rapid drying does no good. Even in turf the soil crusts over so that grass growth is held back. Where ground is worked the nubs exposed by cultivation soon become impervious to it. Meanwhile the wheel ruts, made in a time of wetness, are set into fossils of another era by wind and sunshine. The lurching of the tractor as it passes across them is one of the few reminders of the weather that was.

Where not saturated the grass grew steadily through most of the mild winter, but it wasn't to last. The wind that set in from the east and northeast around the equinox reached a particular ferocity at the end of March and for two or three days carried frost with it. One unwelcome effect was on the grassland where the tips of the blades were scorched brown while the body of the grass took on, for a few days, an odd purplish cast hitherto unseen. In the field most exposed to this effect an oak had earlier been brought down from its hedge by Storm Desmond (or some-such). Clearing up the resulting mess in early April we were surprised to see the contrast between the bulk of the discoloured grass and the still fresh and soft green of that little portion of it that lay protected in the lee of the fallen trunk.

The grass now is nothing to crow about, though it did at last get enough of a wetting (at the end of April) to freshen it up a little. The cattle, released from their winter sojourn in the shed, are getting through the old pastures sooner than we'd hoped while the other fields, currently shut up for silage, and which were formerly the decent meadow fescue and timothy leys which replaced our field vegetables, are now over-populated with soft brome or lop grass, a species of no agricultural worth whatsoever. For all that it is a pretty grass in flower, with plump glossy heads that shimmer in the sunlight as the wind sets it dancing, it produces almost nothing in the way of palatable leaf. In that respect it is a depressing sight, at least to the farmer.

The shift from wet to dry soon had the green woodpecker shouting for rain. In this it echoed my own anxiety. Perhaps because I never got over the drought of 1976 I'd sooner have six months of incessant rainfall than six weeks without any, and even a fortnight's dry is enough to get me twitchy. This bird is known also as the Yaffle from its distinctive laughing call, and as the Rainbird from its supposed ability as a forecaster. There is no doubt that it has an association with rainfall, but whether as herald or supplicant I'm not sure. For the two weeks either side of Easter it kept shouting, so much so that it began to get on my nerves. A third of an inch on the 17th, helpful to the veg ground but scarcely penetrating the surface under grass, quietened it down for a couple of days. It was soon at it again, but then had more of a rest when the end of the month at last produced something useful for the rain gauge to measure. Since then it has been fairly quiet, but if its job is to summon rain it needs to get back to it.

As winter turned to spring I've kept an eye on the ash trees. It was obvious enough by the end of last summer that the die-back disease was beginning to take a toll of the young growth of coppice shoots and of individual seedlings. Less obvious was what was going on with the mature trees whose young wood is out of close sight. Now that most of the ash buds have opened (last to leaf and first to shed them) it can be seen that some of the infected young shoots are dead, but that many, while losing their top-growth, are budding strongly from lower down the stem. The hedgerow trees still seem largely unaffected, though a few – mainly the younger ones – are showing signs of damage.

Elm disease set about altering the face of much of the lowland English landscape more than fifty years ago. Within ten years most elms that grew outside woods and were above fifteen feet in height had died. There were half a dozen lofty corpses still standing on this farm when we came here in 1984. Yet there is still plenty of elm here. We have sections of hedge that are thick with it, still full of vigour. The disease seldom kills the roots and it seems that suckers will continue to spring from these for ever and a day. Once they reach a foot or so in girth they attract the bark beetle and the fungus that it carries, become infected in turn, and wither and die. There is nothing new here – outbreaks of elm disease can be traced back through the written record and through the evidence of annual rings seen in old timbers. The pollen record shows that in the years around 3800 BC elms across the whole of Northern Europe suffered a similarly massive decline to the one we have witnessed.

Elm's strength is that it continually sprouts from the root; its vulnerability is that being clonal it doesn't have the genetic diversity to circumvent the disease. Ash can only reproduce through seed. Its most hopeful defence lies in its diversity, which is well shown by the great variation in the timing of bud-burst of different trees.

The early morning of the 12th of May saw a frost that burnt off the tops of our potatoes and the opening leaves of the walnuts too. More helpfully it withered the emerging fronds of bracken, small patches of which occur on our steeper and rougher ground. It also blackened the newly sprouted tips of the more exposed and still living ash seedlings. Whatever next?

NATURE NOTE 45

Seed sense

Autumn 2020

David Frost (Organic Grower 51) is right – irrigation often enough induces rainfall. What it's not so good at is making weed seeds germinate, or so I've found. I once worked down a plot, known to be full of weed seeds, into a decent false seedbed – and then, as the weather was set dry, gave it a soaking. Some weeds appeared of course, enough to make me feel I'd achieved something when I rubbed them out before starting to plant up. But sometime later there was a day's rain and after that the weeds came up in earnest, the emerging seedlings jostling their way through the surface, a tide of green that made mockery of my attempt to germinate them with the aid of piped water.

I've never tried to artificially germinate a false seedbed again, but have often observed since that moisture alone does not necessarily produce the flush of weeds you might expect. In some circumstances weed seeds remain mostly unmoved even by quite heavy rainfall. Thundery downpours, unless they signal the start of a wet spell, have only a limited influence on weed germination. I remember in particular one good soaking in a dry time. Interested to gauge its effect I marked the tip of a stem of buttercup squash. Within 24 hours it had extended by nine inches, but the weed seed remained mostly unimpressed. Even a few wet days together may not be enough. An example of this came at the end of April this year. It was, for most of us anyway, a dry spring. In March the usual flush of early weed followed the months of rainfall. After that we had frost and unusual heat by turns, but no more rain for weeks – and no new weeds either. Right at the end of April (naturally just as the cattle went out to grass) three wet days produced nearly an inch and a half of rainfall – enough surely to pretty much make up the soil-moisture deficit on cultivated ground. I confidently awaited another flush of seedlings, but it didn't come.

The predominant weeds on our reduced bit of vegetable ground are chickweed, speedwell and annual meadow grass, with dandelions drifting in from their harbourage along the base of the deer fence. Few of these or anything else much made an appearance then. What did come up was a lot of fat hen or lamb's tongue, *Chenopodium album*, together with its fresher green and non-mealy cousin goosefoot (aka allseed), *C. polyspermum*. This family are late spring and summer germinators in the main. They are not frost hardy but revel in heat and drought, emerging and growing steadily at times when other species remain unmoved. My conclusion – drawn from the presence of these and the absence of any other species – that we were in for a summer-long and god-awful drought, happily turned out to be mistaken. All the same the next six weeks were almost rain-less, and towards the end the days became excessively hot. The goosefoot would have had a fine time of it, had not the drought and the smallness of the plot made it easy enough to weed most of it out.

The heat broke down in early June and not long after we had 2½ inches of much-needed rain over three or four days. This time all sorts of weeds germinated in vast numbers. It seems that somehow their seeds sensed more damp weather and cool weeks to come, allowing time enough to establish and set seed before drought and heat returned, as they eventually did.

There's something going on here, but I don't know what it is!

Seeds of perennials, herbaceous or woody, can be very choosy as to when they decide to germinate whereas annuals and biennials, by their nature, need to seize the time each season allows. It's true that in some species a proportion of the seed produced will have an inbuilt dormancy, an insurance against lean times. Cultivating ground here that hadn't seen a plough in living memory our annual weeds (just for that season) consisted of an occasional and solitary fat hen or black bindweed. Instead, we had creeping thistle to occupy us which doesn't need seeds to assure its future – the roots are enough. On the whole though the seeds of cultivated plants will come up if they are close enough to the surface to hazard it. This doesn't need rainfall – tap water will do. We sow our food crops and our flowers into substrate or soil, in modules and tunnel beds and the open ground, with the reasonable expectation that given sufficient moisture they'll soon emerge no matter what conditions await them. No doubt if we sowed modules with chickweed there wouldn't be that many blanks. In the polytunnel it's not as if weeds find any difficulty in perpetuating themselves with just a lick of dampness from the seep-hose, never mind a soaking from the overhead.

And yet out in the open ground and under the sky some selectiveness is at work – even some foresight of the weather to come – that determines whether or not the seed bank of the soil will pay out in spades. What signals are given? You can see that there are differences between irrigation water and natural rainfall. Rain contains small amounts of nitrogen. It is likely to be closer to the ambient temperature than cold piped water. Usually it is delivered more slowly and with a lighter or at any rate a different rhythm and touch. One way and another it is conceivable that rainfall can have a more stimulating effect on seeds than the mechanical application of water, but this goes no way towards explaining the apparent ability of weed seeds, even when rain has wetted the ground, to choose their time depending on the weather to come.

False seedbed – the grower cultivates a bed that a crop could be sown or planted into, but then waits a couple of weeks for the weeds to come up. These are then destroyed with as little disturbance as possible, after which the crop is put in. So long as the weather co-operates it's a technique that can markedly reduce time spent on further weed control.

NATURE NOTE 46

Plant life

Winter 2020

There is another thing about weed seedlings, and that is their apparent ability to germinate preferentially among the very crop plants that they resemble. I can't be the only organic grower to have come up against this notion, while shuffling up the row with eyes and nose to the ground. I had a reminder of this the other day while weeding out a newly planted flower border. Over the several square yards of the whole I came across only one seedling mayweed. It was hidden among the prostrate fronds of a little anthemis for which its leaf shape was a spitting image, save for a difference in shade between them. Probably mayweed is always a likely candidate for this sort of subterfuge. That's because it's the form of the carrot that it most commonly resembles, and of all crops it's carrots that are most likely to demand hand-weeding and thus give the greatest chance for this sort of observation, or fancy. Alliums are another crop that often want close attention in their slow infancy. Then you find it's tiny grasses that have seized their chance to trick the eye. Add to this the belief I've come across among weeders that "it's only grass" and can be ignored and you have the beginning of a problem.

It's all illusory of course, simply a result of more of the seedlings that happen to resemble the crop being overlooked than those that don't. All the same I wouldn't put it past them – plants I mean. It's not as if they are not alive, and all things that live use whatever strategies are available to hang on to life, to flourish and increase if they can and at the least to perpetuate their kind.

It's one of the pleasures of being an organic grower that you do often have your nose to the ground, this being the position from which life in all its miniaturised fascination is best observed. The livestock farmer and still more the arable farmer, especially now that they are cocooned in a sealed tractor cab, see little of this – though they may have more familiarity with the longer view. It's a truism that farmers on the whole make lousy gardeners. Broad acres of cereals or rape don't encourage interest in the individual plant while the stockman, used to a charge that moves and makes noises, tend to view plants as inanimate objects.

But the art of good husbandry, stockmanship – the observation, empathy and care that together make a good farmer – applies just as much to plants as it does to animals. In some ways more, as plants cannot move to find shade or water – something to explain to your planning officer perhaps. The crops we grow are as much a product of domestication as the animals we keep, and as we have domesticated ourselves in the process. They have swapped the determined survivability of their wild forebears for a fat and cosseted life, and that's why we have to nurture them as they succour us. Or is it the other way round? Either way our crops and animals would be as lost without us as we'd be lost without them. We're all domesticated together.

It's different in the wild. Our weeds can't move of their own accord, but they have ways of spreading their progeny about and generally making a mess of what we're trying to do. You could call it agency, without that implying any more than a mechanical and choice-less mode of life. But at the other extreme of growth – a tree that has stood for several human lifetimes and that, as it towers above us, exhibits clear signs of individuality – you might believe that it knows something we don't. The branches reach into the sky we cannot ourselves inhabit and the roots go places where we can't follow. I remember reading a reference to Native American forestry practice, probably in some publication designed to catch the attention of credulous hippies. This maintained that before going into some woodland in order to fell a tree the tribespeople would announce their intention of so doing. At this all the trees would faint. Quite what the point of this was I forget, presumably a form of self-anaesthesia, but the idea of trees communicating didn't seem far-fetched. Trees and all other plants form communities, and communities communicate – of course they do!

Our monotheisms proclaim humanity's separation from the rest of life as the Crown of Creation, empowered to do what it will with the world. The Reformation and Enlightenment did for the remaining magic and since the Industrial Revolution, when most of us left the land, we've practically ceased to invest different plants and animals with special powers and qualities. Our sciences have tended to reduce everything to separate bits of disconnected matter, as knowledge has replaced wisdom and been replaced in its turn by information. Lately though the process seems in some respects to have got to the other side, microbiology revealing complexities and relationships that previously we could hardly have guessed at. I think of Peter Wohlleben and his book 'The Hidden Life of Trees' which has had a wide readership and which credibly animates the woods and forests. Among other things he shows how trees do communicate, through scents and via the mycorrhizal network, which he calls the wood-wide web. And if trees – why not all plant life? It's only a matter of our perception of size, and as Chairman Mao put it in some other context – "size is of the least importance, for a giant corpse only feeds more vultures".

Here we come back to organic agriculture, which time and again has shown itself in advance of the mainstream in understanding and wisdom. It was Sir Albert Howard in 'An Agricultural Testament' (1940) who first gave mycorrhizal associations the importance that they are now universally recognised to have. Lady Eve Balfour built on and expanded this theme in 'The Living Soil' three years later. Both are still worth reading, as much now as they were then. Meanwhile you can now buy mycorrhizal fungi from your garden supplier – £11.95 for a 360 g pouch (probably cheaper on the internet). But you're an organic grower, so it's highly unlikely that you'll need to. The fungi will be there already.

While hedging

Spring 2021

There's something over 2½ miles of hedgebank on our thirty acres, together enclosing ten fields and two orchards. As well as increasing the banks' capacity for shelter, the hedges that clothe them were productive enough to explain why a farmer with 24 acres could afford to lose 2 of those acres (at a conservative estimate) to field boundaries. They were more than just repositories for cleared rocks and passive enclosers of plots of land – cut rotationally they were an essential part of the farm economy, answering its needs for construction, fencing and firewood. A range of species is present but hazel predominates, often as pure stands over the course of many yards. Helpfully hazel doesn't attain much over twenty feet and is a natural coppicer. It was favoured for its production of long, straight rods which were easily split or used in the round for a variety of purposes. Now we think of bean poles and props for washing lines, with the branches making pea sticks, but these were minor uses compared to hurdles (made in industrial quantities before the advent of wire fencing) and wattles, which – combined with daub (i.e. mud) – made the durable and ecologically sound precursor of plasterboard. A particular property of hazel is that the stems can be twisted to part the fibres and then bent to extreme angles to make thatching spars and knotted around the faggots (compressed bundles of small sticks) that were fuel for the bread oven. Before grey squirrels you could also reckon on a good harvest of nuts.

By the time we found it the original farm had been grazed as one for a good few years, with cattle pleasing themselves as to whether they used the gateways or made some other route more to their liking. The hedges had meanwhile seen no maintenance for several decades – tree ring evidence suggested about sixty years.

The immediate priority was to first get the hedges back to the line of the banks – removing suckering blackthorn and the plants that had slumped outwards – and then to let more light and especially air into the vegetable ground. Given the labour involved and the narrow seasonal gap available for such work this was a slow process and barely completed by the time we ceased commercial growing nearly thirty years later (and there is still a fair way to go on the periphery). Some of this was achieved by simple coppicing, giving rise to rejuvenated growth that could later be laid or, if needs must, kept in order with a hedge trimmer. A lot of it we did lay – "steeping" as it's known in these parts – from the start. With care this works well enough even on hedges with big old stems. You cut most of the way through the chosen stem as close to the plant's base as possible, leaving a strip of bark and sapwood to act as a living hinge. You then lower it down so that it rests a little above the horizontal along the line of the hedge. The stems can be pegged in place if necessary with crooks cut from the discards, or if you're not fussy you can tie them in with bale cord. Removing much of the now prone top growth reduces demand on the flow of life through the hinge. It's engrossing work, particularly when the thicket of stems have got to their mature height and developed a fair-sized crown of tangled branches. You're up on a bank with a juddering chainsaw, trying not to catch the chain on a stone that's escaped your notice, and then before it decides to fall of its own accord you've got to get the stem down gently enough that the hinge doesn't break. A good old hazel stool may have twenty or more substantial stems to it besides smaller ones, and with little gap between them at the base, so that it's a puzzle to know what you can keep and what you must remove.

When it's done there's a new world revealed – a close view exposed and a distant view where there wasn't one before. The trampled earth soon heals and by the following spring the bank shows a flush of wild flowers, now released from shade. Early in the first summer vigorous shoots spring up from the cut stool and from the numerous dormant buds along the steepers. The circularity of the hedge's life is ensured.

Lately I've been doing a bit to a hedge last cut a dozen years ago – the short interval making for easier work. It's also been appropriate to my years in that it grows on the flat and so doesn't require too much athleticism – at least on the side from which I did most of the work (on the other is a six foot drop). One lovely afternoon in January just at the start of a cold spell, when we'd lost the sunlight here but it still shone on the further side of the river valley, I was squinting up through the stems into the confusion of branches above, figuring out my next move, when into the corner of my sight came two large birds, high, high up and white against blue, pursuing an undeviating course down the valley from north to south, toward the distant salt water. They flew close and parallel, one a little behind the other. It was hard to make them out – later I thought they had perhaps been swans, Bewicks maybe. All I could see of them, their bellies and the sweep of their wings, was white – quite brilliantly so in the brightness in which they still flew. I was following their progress southward, already entranced, when suddenly they ceased to be two diminishing white shapes and became flashing stars. It was an extraordinary moment, white becoming light. The sun's rays were now reflecting off every beat of their wings. I watched the two stars flashing side by side into the distance until lost in the haze of the horizon. Such moments of serendipity, or even of transcendence, remind us (should we need it) why it is that we prefer to spend our lives under the sky.

NATURE NOTE 48

Dandelions

Summer 2021

It's hard for an amateur who's not in the business of measuring data to be sure of changes to the living ecology of a place. They tend to be in the nature of subtle variations to a complex pattern, accumulating over time. It's often only when such variations have become cumulative that you wake up and register that something is different – trees where there were once just bushes, or a patch of brambles that has swallowed more ground, seemingly while you've been sleeping.

Lately though there has been an explosion in the dandelion population, here and on neighbouring ground at least, which would be hard not to notice. I suspect that it is due to the arrival of one or more alien strains. Somewhere along the way dandelions abandoned sexual reproduction, the embryo instead being formed in the seed without fertilisation, a process called apomixis. Through mutation this has resulted in the development of hundreds of micro-species, each one a collective of genetically identical clones. Mutations that disadvantage the plant do not persist, but when a micro-species possessed of a competitive advantage arises it can spread with a single-mindedness greater than is found in species that rely on sexual reproduction. The increase is especially evident in those bits of the permanent pastures where badgers have been exposing bare soil while ripping up the species-rich turf in their search for grubs – don't they know that it's a County Wildlife Site? But this is nothing to the change brought to the leys (if I can still call them that) which replaced the vegetables in our more level fields. Here the flower has taken on what was formerly the prerogative of buttercups, dominating the sward so that for a few weeks in late April and May green is subsumed by yellow, a herald and signifier of the coming of May, never previously one of its traditional aspects. The buttercups still have their day but are now accompanied in it by the ghostly globes of the ebbing dandelion tide.

On the rough margins too they now wax in ever greater numbers. There's not much concrete here and the shaggy edges of the hardcore that we make do with instead has always provided a home for ruderals – plants that colonise waste places (literally, rubble). Among these are plantains, attenuated docks, annual meadow grass and so on. Dandelions have always been part of that company of course but it's only lately that through weight of numbers they've started to make a bit of a nuisance of themselves. The front apron of our packhouse shed is where I've always filled the seed module trays. You'd think by my age I could have come up with a means that didn't require me to do the job on my knees, but still it suits me well enough. Now though I really should get up off the ground, not because of my knees but because of the numbers of dandelion seeds that, on a dry day, come scudding across the floor to fetch up at that very spot where they can mingle with the Klasmann substrate.

An average globe holds 180 seeds, each one an eager hang-glider. You can buy these seeds if you want to. The internet is thick with sites, but best to go to Jelitto who will sell you cleaned organic *Taraxacum officinale* seed at 2 € for a gram, 248 € a kilo (124 € if non-organic). There are around 1,200 seeds to the gram, but Jelitto recommends sowing 2 grams to be sure of a thousand plants. I can't imagine buying a gram or come to that a packet – the equivalent to the contents of which were stuck to my gumboots when I came in from looking at the cattle this wet and soggy morning. As well as being a good nectar source for insects the dandelion does have its human uses – among them Dandelion and Burdock and an inadequate coffee (roots), Taraxagum

– a rubber alternative (sap), salads (leaves) and Biodynamic preparation 506 (flowers). In the unlikely event that you'd want to drill it then it is something to have the seed cleaned of its feathery pappus. But for most of us it doesn't need any more assistance in getting about the place than we already unwittingly provide.

Like a dock, dandelion roots soon become close to indestructible. Unlike a dock, which reproduces only from the top few inches, all parts of its root are capable of regrowing and even if you spud one out to a depth of six inches it will reappear sooner or later. The leaves too are tough. Their upper surface grows a little quicker than the underside enabling them to lie tight to the ground, suppressing feebler plants. It also has strongly contractile roots, commonly found in bulbous plants, that pull down on the crown to keep it just at the soil surface. In vigorous grassland it is not quite the weed that you might expect because here the leaves are held up at an angle, not much supressing the grasses, and both leaves and flowers are eaten by stock. It was never much of a nuisance in tractor-worked vegetable fields, but now I find myself in both polytunnel and garden sometimes digging out plants that have hidden themselves away for long enough to get a proper grip on life.

It's odd that we should know a common plant by a French name, itself derived from mediaeval Latin – dens leonis, presumably in reference to its leaves. Its old vernacular names refer to everything but the leaves – for instance 'golden suns' and 'yellow gowan' for its flowers, 'fairy clocks', 'farmer's clock' and 'tell-time' for its seed-heads, 'monk's head' for the bare receptacle that's left when the seeds are gone, 'devil's milk-pail' for its sap and 'piss-a-bed' for its diuretic properties. Not that 'dandelion' is a bad name. Read as English – dandy lion, rather than lion's teeth – it's actually rather a good one for these little ramping lions of the sun. It's a nicely affectionate one too for a flower which has such an appeal to the very young in its simplicity and brightness. Here's a place where, as the poet Kenneth Patchen has it, "the sun spends his fabulous money".

Adam/Had 'em

Autumn 2021

The above may be the shortest poem in the English language (a contender is "Thorp's/Corpse", inscribed on a miser's headstone) and is taken to refer to microbes, or perhaps fleas – the original intent seems unclear. As a good proportion of the cells in a human body are not in themselves human, but are fellow-travelling bacteria etc., it's fair to assume that Adam did have microbes, even in Paradise, and if Adam – so with Eve. Fleas though? That goes against the grain. Some might say that all creatures deserve a paradise (which in the flea's case would need to be stocked with sentient flesh), but perhaps these and their blood-thirsty fellows only came with the Fall.

I don't know about Paradise, obviously. Its defining aspect is that no one lives there, so we'll leave such questions to the theologians. At any rate the likelihood is that if Adam was bitten Eve wasn't, and vice versa. That seems to be the way it is, with fleas in particular and with biting insects in general – they show some discrimination and exhibit unflattering preferences when choosing their targets. When an old cat of ours died, years ago now, it was to me alone that it bequeathed its fleas. This was a bequest I could well have done without. They lingered on, not thriving by any means but irritating all the same, for two or three months. I found it expedient to remove and shake out my clothing each night while standing in the empty bath. The displaced fleas showed up against the white enamel and could get no purchase on its smooth surface. They could then be flushed down the drain and into oblivion, despite being capable of a quite nifty sidestroke.

There's something sordid and depressing about being flea-ridden – it's no joke! – and I can imagine that to be infested with fit and active human fleas (rather than a few stray cat fleas) would be severely debilitating. It and the other blood-feeding parasites – bed bugs and body lice, which are entirely dependent on the body of the host – have no particular association with life on the land, far from it, and indeed we tend to think of such a life as inherently healthy. But here on earth, paradise lost, the country worker has to put up with a variety of other generally winged and more or less irritating life-forms that need some measure of animal blood. This is most evident in the summer months, and the grower presents an easy and often slow-moving target. It's not surprising perhaps that through the winter – as we contend with mud and frost, chapped hands and chilblained toes – we look forward to the pleasures that the warmth of summer brings. But when summer comes the biters and itchers sooner or later remind us of their existence, and then we remember that winter has its benefits too.

Sometimes it's obvious that you've noticed a cleg or a mosquito too late and must put up with the consequences, and it's impossible to miss midges in their unstoppably irritating hordes. Often though there's no identifiable cause to the various itchy bumps and bites that appear on the skin during days of summer spent out on the land. Whatever the cause distilled witch hazel has a soothing effect and is cheap and cheerful. There used not to be much to worry about with the biters native to this country beyond some discomfort, no chiggers or tsetse flies for instance. Lately though the explosion in the deer population has induced an equal explosion in the population of ticks, and these little arachnid mites are vectors of Lyme Disease, a chronic and multi-faceted affliction which is difficult to treat. Ticks used to be not uncommon on dogs and farm livestock but they seldom bothered humans. Now they are everywhere – earlier this summer I found four on myself one day and three the next, but apparently this is nothing compared to what forestry workers have to put up with. Not all ticks carry the disease and even if they do they need a couple of days in which to pass it on. Although tick bites can itch and may do so over a protracted on again/off again period, they don't always do so – so keep an eye out for a little raised spot with a black speck at its centre. During the tick season any bull's eye shaped rash or inexplicable flu-like symptoms are worth a visit to the doctor. A course of antibiotics, taken soon enough, should prevent the disease developing further.

I used to think that all ticks were fat and obvious, like plump sultanas. If they are you can remove them with pullers designed for the purpose, but these are engorged specimens, inflated with the host's blood. Prior to that your tick may be no bigger than the full stop at the end of this sentence and needs to be removed with (preferably fine-nosed) tweezers. Un-engorged even the adults are only as big as a sesame seed. The tick's lifecycle is long and fraught with hazards. Its course from larva through nymph to egg-laying adult occupies three years at least. Determined and tenacious creatures, they can survive months and even up to a couple of years without feeding.

Viewed through a hand-lens the dot of matter responsible for all this aggravation resolves itself into a not unimpressive crab-like creature with its mouth parts buried in your flesh and its bum in the air. It must be the deer that have brought about the change in numbers, that's the rural consensus, though the British Deer Society (patron Fr. Christmas?) denies it and seeks to shift the blame onto birds instead. But the tick population has waxed mightily alongside that of the deer, whereas over the same years the bird-life (leaving wood pigeons aside) has been all in the other direction. We never had deer and only saw ticks on bullocks or the cat. Now the deer are ubiquitous and so are the ticks. Early summer is their prime season, but they are active in autumn and into winter too – I found my last tick of last year at the end of December. The associated irritation comes and goes but can persist for weeks – probably because the mouth parts have been left behind. They are tough little buggers, and alive or seemingly dead are safest when they too are flushed down the drain.

The common British species of tick (and the one that carries Lyme Disease) is Ixodes ricinus – for which sheep-, deer- and castor bean-tick are synonyms. Full information is available from the UK Health Security Agency's Tick Aware toolkit, easily found on the internet.

Survivors

Winter 2022

Early in my farming career there was a play about docks on Radio 3 which both Jan and I remember to this day. It had some effect on the way we viewed humanity's relationship with the land. I say "about docks" but actually it was the docks' play. Called 'On a Day in Summer in a Garden' and written by Don Haworth, the cast consists of three docks, a grandfather and two youngsters, which (for the benefit of the listener) communicate in English. There's a man and a woman too, but what little they say is what the docks hear – which of course is gibberish. The man has a knapsack sprayer. The docks had been spending a balmy afternoon listening to grandfather's tales of how they had taken over the garden, generation by generation, but now they are drenched with burning poison which tears at their roots. Through his agony grandad exhorts them to hold on, to cling to the earth and stone. One cannot, blackens and dies. The man meanwhile has poisoned himself! He froths at the mouth and expires. Rain falls, it washes the fear and pain from the remaining two docks. They live to fight another day.

Docks are fighters and we need hold no sympathy for them – not that we'd want to poison them. Their only evident virtues are in mining minerals and calming nettle rash. I'd sooner have chicory doing the first job and for the second – the docks can grow where the nettles are. Still, when it comes to persistence and assertiveness, you have to hand it to them.

There's another dock story, more of a vignette really. Henry Williamson is best known for 'Tarka the Otter' but he was also a sensitive observer of human life. This little tale of his concerns an old man and his battle with the docks that infest the cobbled path leading to the door of his cottage. After constantly poking and pulling out what he can get at he's just about got on top of them, so that in what turns out to be his last autumn there's nothing of a dock to be seen above ground. That winter he dies. Spring comes, the days lengthen out and as they do so – here and there a shoot emerges. He's gone, the docks have survived.

As the years go on this story gets closer to the heart. It comes to mind as I fork out the stroyle (or couch) that finds its way to come creeping into the polytunnel. I think of it too when I see the chickweed (for instance), by no means reduced in its potential after getting on for forty years of false seed beds, flaming, hoeing and weeding. As for docks – I'm sorry to say that a legacy of our tenure of this land will be a greatly increased stock of the things. There weren't many on the farm when we found it, and they never became a feature of our vegetable ground. It was when the vegetable growing contracted and we started keeping cattle of our own, instead of renting out the summer grazing, that the docks arrived en masse. The two events seemed oddly coterminous, as if the mere fact of our becoming livestock keepers was enough to give the docks a following wind. In truth though, the seed was in the organic straw that we bought in for bedding. Once the seed is in the muck it's back in the fields, no matter how carefully you turn the heap. From there it's in the silage, then in the shed and so back in the muck again, an unbreakable cycle. It's not long before one man, armed with one Lazy Dog dock digger, can only look at the result and sigh hopelessly.

Docks, stroyle, chickweed and the rest of them – they'll all still be here when I am not. I'm happy to believe that this will be so. And it is not just through bullies such as they that life continually asserts itself. I let a leak in the plumbing in the shed go on too long before I got around to fixing it, and before I did a band of moisture had extended a few feet along a sort of redundant gutter at the base of the inside of the block wall. In no time a vivid green patch appeared which on closer inspection proved to be a thicket of seedlings – vascular plants rather than the liverworts or perhaps moss that you might expect in that situation. What they are I don't know, but I've potted up a bit of the mat of young leaves and roots and I may find out. How they got there, this swarm of individual plants all in a mass, is a mystery too, but what I find most arresting is their ability to thrive on a barely visible film of substrate composed of shed dust and the odd dead insect fallen from the window above.

It's not just plants that will find their place. There's a beetle for just about every situation – one that lives on museum specimens, another that can get by on smelling salts, and so on. There's the petroleum fly whose larval development takes place in pools of spilt or naturally occurring crude oil (it feeds on other insects trapped in the viscosity). Then when it comes to fungi, never mind bacteria, it begins to become apparent that there is no end to their ability to survive and prosper in places where we humans cannot go and can barely imagine. All this is the saving grace of the situation that humanity finds itself in and a consolation in the yawning chasm of our own self-destructive mortality. Let's not talk about saving the planet. The life of the planet will endure without our help. To think otherwise is just to go on repeating the same ignorant hubris that has got us where we are. Biology – or Nature if you will – is the boss, the one god (if any) that warrants our obeisance. But yes, of course, we have to look to saving ourselves.

Those plants turned out to be ferns, as I might have guessed, and not of one species either. I now have two male ferns and a hart's tongue sharing one pot.

Epilogue: Two Valleys and a Farm

The farm sits on a shelf of land formed by the interaction of two watercourses and their valleys. The greater valley is that of the Teign, which runs due south about half a mile beyond the lowest point of the farm. The smaller, that of the east-flowing brook, marks the boundary of the farm's land all along its lower edge.

The Teign gathers its headwaters from the north-eastern quarter of Dartmoor. Ignoring the easy route to the sea offered by the Sticklepath Fault, a wide south-running trough between high land on either side, it instead carries on straight eastward, chiselling a course for itself through the hills in front of it to make a gorge several miles long and for some of that length over five hundred feet deep. Although somewhat disfigured by dark stands of conifers, planted and displacing the native oaks in the 1930s, it is a place of spectacular scenery and much beauty. Emerging from its gorge the river turns to the south and enters the Middle Teign Valley, generally just known as the Teign Valley or – to locals – simply The Valley. Here, the hills standing back on either side, it flows past three or four parishes on each bank, before making the last few miles of its course to the sea through the Bovey Basin, along the line of that same fault which it had earlier ignored.

This valley has more than local significance. The ridge of Haldon that marks its further edge can be seen from much of Devon east of the Exe. Approaching Exeter from that direction it looms darkly like a rampart, and if you want to travel further west a stiff climb can only be avoided by following a circuitous course along the coast. This rampart marks the frontier between England's two foundational zones – the Highland and the Lowland. East of this line is the England of more recent geological time, of lesser elevation with less rainfall, and of richer soils. West of it the country lies on older impermeable rocks, with higher, wetter hills and less fertile soil. This is a generalisation of course – there's better and worse land on both sides of the divide – but (as has been aptly said) it's generalisations that make the difference between the Beatitudes and a telephone directory or user's manual, and the distinction is a real one. The frontier is taken to run from the Exe to the Tees, and though not without gaps and inconsistencies elsewhere there is no doubt that the Haldon Hills mark its southern end. It is reinforced by the secondary rampart formed by the valley of the Teign, lying deep below those hills and then rising again on its western side in a steep escarpment above which is the easternmost of the Dartmoor granite. This further tableland drains into the Teign and is mostly contained within parishes that border the river, and so to that extent may be considered part of the valley itself. It is somewhat higher than the Haldon ridge and from it you can gaze across into Dartmoor proper, with its bare hills and rocky summits.

The different conditions on each side of the zonal frontier have had a formative effect on the history and economy of the two zones. You can see this effect writ large as you cross the watershed from the Teign to the Exe. Ahead are deep red soils and gentler gradients underlying larger fields, and bigger farmhouses and farmyards to go with them; behind are brown earths more thinly spread over rock or clay. The land here is an irregular patchwork of small fields and small woods clothing the spurs and combes that run out from the heights on either hand. It's later to start growth in the spring because often poorly drained, and therefore slower to dry out and warm after the winter's rains. Away from the granite plateau to the west there is little ground that does not exceed the 12° of slope where arable cropping becomes more difficult and more costly, even if the nature of the soil will support it. Once it gets going grass grows well enough here, but the scope for productive farming is more limited than on the other side of the watershed.

Those of us that live here consider the valley to be a place apart and distinct in itself. Its scenery achieves a kind of perfection, to which the many dispersed old farmhouses (that seem to have arisen from the earth) and the rambling hedge-banked lanes that connect them give a sense of the easy passage of time which is itself a kind of timelessness. Perhaps the only thing lacking is the ruins of a mediaeval castle – or the mound of a motte would do. But the valley has never been on the way to anywhere, nor (it seems) has it been the place of great events. The English Civil War was the cause of a rumpus or two locally, but few parts of the country avoided some upset during that miserable conflict. Then in the nineteenth century there was a brief and largely still-born mining boom, exploiting the silver-lead lode that shadows the edge of the granite. Of that only some heaps of waste remain, slowly hidden by reasserting greenery. It's in the fields and farms that the valley's true history is found.

*   *   *   *

The smaller valley, one that can claim no more than local significance, is that of the brook. It rises at close to 1000 ft in a patch of wet and overgrown ground a few hundred yards below Heltor Rock, now since the demise of Scattor Rock the easternmost tor of Dartmoor [see note 39]. It has to drop 800 ft in two and a half miles to get down to the level of the river, so it is in some haste for all its length. Once it gets under way the banks on either side are steep, its course deepening as it flows beneath our neighbouring village, perched on its hilltop. Then as it comes down through the line of the scarp it cuts a deep cleave (the site of Scattor is several hundred feet above its right bank). This is the shortest route between our village and its neighbour, but many people prefer to drive the long way round. The ascent is no less steep, but that way the road is dry and not constantly rutted and pot-holed by the water that runs down and across the tarmac, nibbling at its edges, as it seeks a way into the brook.

The brook makes almost all of its course through a thick fringe of trees, or through actual woodland. Upstream it is flowing over granite, and between the boulders its bed there are patches of bright, gravel and sand in its bed. Lower down and through the cleave its fall is too rapid to allow much accumulation of smaller particles. Instead it rushes headlong in a succession of falls and cascades over and around the great lumps of stone that litter its way. Here it transects the band of the metamorphic aureole, where aeons ago the original rock was transformed though the intense heat of the granite intrusion, deep beneath the earth, into seamless elvan, very much harder and heavier than the granite that neighbours it. It's darker too, the stones in its bed coated with a black patina deriving from the presence of manganese in the water.

Beyond the cleave the escarpment marches due south and away from the brook, with the land in front of it falling relatively gently towards the river. The interaction of the two valleys, the greater and the lesser, has created a terrace of level land (level at least by local standards) which begins at the inner margin of our uppermost field and extends through the rest of the farm, widening as it goes. Below it, all the way, is a short but steep drop to the brook, above it some rising ground that is comfortably within the critical 12° of gradient for arable cropping. On its further side the brook is flanked by a gradually declining spur that runs out almost to the river. It was once mostly cleared for grazing and enclosed with banks, but its steepness makes it a wonder that this was ever done [note 36]. Now clothed by woodland its height shuts out any view to the north, and shuts out those winds that don't find their way down the cleave.

*   *   *   *

It might be thought that less productive land would give rise to bigger farms by way of compensation. In fact the opposite applies – to him who hath shall be given. A prosperous farmer can more easily acquire further land and can afford to employ the labour to work it. By the same token there is some truth in the saying that the less land you have the harder you have to work. Unless considering really poor ground of upland grazing and extensive livestock the farms of the Highland Zone tend to be on the small side. Managed chiefly by family labour, historically they had more of the nature of subsistence farming about them than those of the gentler valleys and plains. Our own farm is a case in point, and rather an extreme one at that as it seems never to have exceeded the couple of dozen acres that we purchased in 1984. That it had, even so, supported a family in some degree of comfort over several centuries was evident from the ruined mediaeval farmhouse thrusting up oak crucks and fragments of cob and stone walls above the encroaching vegetation. The still standing granite ashlar fireback and the once smart oak stud-and-panel screen, that together formed the house's cross-passage, show that the farm's tenants were able to keep up with sixteenth century fashion as well as did their more generously endowed neighbours.

The farm is not far from the village but out of sight of it and right on the edge of the parish. It's reached by a narrow lane, a succession of dizzying twists and turns, that drops down from the public road along the foot of the scarp. The brook marks both its bottom boundary and, at this point, the boundary of the parish too. The unthreatening roar that it makes in its headlong rush as it exits the cleave is a constant accompaniment to life here, gentle and muted through the summer but loud and insistent during a wet time. In the classic view of landscape history the farm must represent the latest stage of mediaeval agricultural advancement into the "waste" beyond the original enclosures of the village and its ribbon of ancient farmsteads. This would place its creation in the twelfth or thirteenth century. It's unlikely to be a matter of clearing thick woodland – Devon seems to have had little more woodland at the time of the Norman Conquest than it does now. Very likely the land had more of the nature of wood pasture resulting from centuries of rough grazing, with patches of scrub and open glades between standing timber. Perhaps more significantly the pioneers here would have had to contend with a superabundance of water, the foot of the scarp being well supplied with springs fed by rain falling on the impermeable rocks above and issuing out below. They also had to contend with a great deal of stone, perhaps more here than in other parts of the parish because of the proximity of Scattor Rock above. Before the freezing and thawing effects of the ice ages this would have been a mighty pile of rock, and a good deal of it ended up littering the land here.

Though enough remained to cause problems to tractor-mounted implements in the late twentieth century, the bulk of the stones was cleared from the land. Rough customers these stones are – often massive and always dense, impossible to work and characterised by awkward angles. Some went into the building of the house and the couple of buildings that made up the yard beside it, though few modern masons would look at them. Others were used for drystone walls – around the yard, along almost all of the brook's course and with a few sections here and there in the field boundaries. Most were deployed in the hedgebanks as a facing to the earth within, though some of these banks were of soil and turf alone. The work must have been done as time from other labours allowed and with no obvious guiding principle. The fields are small (the biggest only three acres), straight lines quite unusual and parallel lines altogether unknown. It is possible though to get some sense of the order in which the work of enclosure was carried out by looking at the map [note 11] and seeing how the hedge of an earlier field cuts awkwardly into the shape of a later one. Meanwhile the water was carefully captured into small watercourses – only a short length of the farm's boundary is not defined by running water. Sometime, probably between about 1600 and 1800, shallow leats were laid out so that water from the little streams could be led into most of the fields on the farm, and then allowed to trickle over the pastures in rotation through the winter to raise the temperature of the soil and thus to stimulate the growth of early season grass.

Even without functioning leats (and without the application of the fertilisers which are their modern equivalent) the farm is notable for its early grass growth. This is sometimes remarked on by those that know about this sort of thing. In the 1870s, not long after the farm ceased to be a self-contained economic unit with a farmer in the farmhouse, its owners – the Misses Addems – were advertising its "luxuriant grass" for rent in the local paper. Why the farm should produce luxuriant grass is a bit of a mystery. The ground is all back-sun, that is – it slopes to the north and east. This is an advantage in a hot summer but, allied with the midwinter sun disappearing into the hill above in the early afternoon, it makes for quite a chilly spot through the winter. The soil is generally of good depth, but being silty in its composition and mostly overlying a clay subsoil it holds a great deal of water, and the evaporation that has to take place for it to dry out is a cooling process. On this account light, sandy soils are naturally sooner to warm up in the spring.

It is a sheltered site – "lew as a box" when it comes to the south and south-westerly winds that we hear overhead but barely feel. The bottom ground, in the hollow of the brook, is in any case intrinically sheltered and elsewhere the smallness of the fields and the hedgebanks with their top growth filter the winds that do get in. These hedges were a great resource when materials for useful poles and timber, and for fuel, needed to be found close at hand. So much of their length being of hazel they also provided a valuable harvest of nuts. The small fields with their hedges and the shape of the land itself – dipping down, levelling out and then falling away steeply to the brook – together make for an extraordinarily diverse and intricate farm. There's no woodland as such, besides little corners of hazel coppice and the sprawl of alders and sallow in the wet bits, but viewed from the farm entrance it's trees that seem to predominate. Walking around for the first time it seemed impossible that all this could be accommodated in the twenty four acres that we originally bought – surely it was fifty at least?

The old farmhouse – cross passage screens, remains of jetty and middle rail

And there was the old farmhouse, the cow house beside it and a small threshing barn (now not much more than a heap of stone) making the further side of the little yard. This is all tucked into the slope at the bottom edge of the entrance field at the point where the extending terrace has begun to make a bit of width. The house is stepped slightly into the bank, and both cow house and barn make use of the differential height to provide ground level access to their lofts. The immediate approach is by a sunken lane which being no more than six feet wide was made with pack-horses rather than wheeled traffic in mind. At the lower limit of the yard is the well – just where you would think the drainings from the cow house might run into it. It has a stone shelter, roughly corbelled, to give it protection. We call it a well and so does the map, but a water diviner dismissed it as a "catch pit". Be that as it may, and it is little more than five feet deep, it provides a unfailing supply of sweet water through the year. Here you might think was the perfect site for a dwelling – snug, unobtrusive and with water to be had a short step from the farmhouse door.

I'd worked on a farm in the north of Scotland, a piece of land cleared of its crofts in the nineteenth century and laid out in business-like fields. The whole of this farm could be fitted into almost any one of that farm's fields, and not unlike a Highland croft, it had provided for a family through a great variety of different enterprises and an intensity of work and attention to detail that made use of all its possible resources. We had a romantic notion that it might provide for a family again.

Exactly when this farm was first established is impossible to determine, but whoever those pioneering hedge-makers were they were not the first to make use of and value this land. Occupation here goes back at least until the second millennium BCE. It's possible that some of the stone cleared from the ground in mediaeval times had formed hut circles, better called round-houses. A few of these are still extant no great distance away on the heights above, where they escaped later enclosure. We can guess at all this because of the number of worked flints that came to light in the course of our cultivations and vegetable growing. These are not just a few stray tools. The finished blades and scrapers are far outnumbered by discarded flakes with among them a few cores – all that was left of the flint nodules when no more potentially usable bits could be struck from them. This suggests a settled existence and a place of work. The finds that speak most of ancient domesticity are a couple of spindle-whorls, the fly wheels that provide hand spinning with its motive force. One in particular is a beautiful object, a piece of probably local dolerite with a regularly carved flange and a perfect hole (to take the spindle) somehow bored through it. The descendants of the spindle tree that might have provided the spindle for it still grow here in some abundance. Lately the deer that surely those people once hunted for their meat and hides and for sinew and bone have made a comeback. Of the other fauna then probably extant I'm not expecting bears and am not persuaded that wolves would be welcome, but the beavers are perhaps not far away.

About the author

Tim Deane was born in Devon, leaving it and returning several times during boyhood and as a young man. Following university and an undistinguished degree in archaeology and history he fell happily into some casual work on a farm. This provided just enough experience to gain him a permanent job on a mixed farm – beef, sheep and arable – in the Scottish Highlands. After two years in the north he returned to Devon and a one-year National Certificate course at agricultural college. This led to a post as stockman/tractor driver on another mixed farm in Cornwall, where he spent six years.

In 1984 Tim, his wife Jan, and their two young children took on a small, abandoned farm in Devon. With no electricity or running water and a tumbled down farmhouse, it was a matter of homesteading while learning the business of organic vegetable growing. Somehow they stayed just solvent selling their produce on the wholesale market but greater security came with the establishment of a vegetable box scheme in 1991. This was the first such scheme to market the entire output of an organic vegetable holding directly to individual households in this way, and was the immediate model for the subsequent expansion of direct marketing with the relative financial security that it provided.

Milton Keynes UK
Ingram Content Group UK Ltd.
UKHW031323180824
447092UK00005B/90